UNIVERSITY OF NORTH CAROLINA
STUDIES IN THE ROMANCE LANGUAGES AND LITERATURES
Number 33

I0651127

MEDIAEVAL LATIN AND FRENCH BESTIARIES

MEDIAEVAL LATIN AND FRENCH BESTIARIES

BY

FLORENCE McCULLOCH

CHAPEL HILL

THE UNIVERSITY OF NORTH CAROLINA PRESS

PREFACE

The Latin *Physiologus* and its enlarged form, the bestiary, are among the best known types of mediaeval didactic literature. They are frequently cited today as examples of serious works of natural history in an age which supposedly relied wholly on tradition from the distant past, and also as illustrations of the naïve credulity of a people who could accept the tale of the capture of the Unicorn as an allegorical representation of the Incarnation. Both statements incline to err in the amount of emphasis and acceptance which they place on each of the two parts of this picturesque compilation — the fabulous description of the real or imaginary animal or bird and the Christian moralization which is derived from it. No collection which repeats the same animal tales in an unchanged form from the earliest centuries of the Christian era down to and, in exceptional cases, even through the Renaissance, can be called anything but a long-lived, uncritical work recording popular tradition. Nor can illustrations used by the Church Fathers to render subtle theological concepts more intelligible and vivid to the unlettered people be presumed to prove that mediaeval man actually believed such examples as were perpetuated in the *Physiologus* and later the bestiary. The present monograph, although treating the second element, the religious, only in a very brief manner, is intended to describe the nature of the contents of the *Physiologus* in the animal realm and to clarify the complicated manuscript tradition.

Few general studies on the *Physiologus* exist in English, and it is hoped that the results of some years of pleasurable research which are presented here will be of help to future students who enter the interesting world of the Phoenix, the Siren, and the watchful Lion. This survey in no way claims definitiveness; it is to be considered more as an interim guide, and its purpose will be unfulfilled if within a short time it is not superseded by a more comprehensive examination of manuscripts and contents. Because of a personal preference for illustrated manuscripts I have largely neglected those barren of pictures of any sort. Although there doubtless remains something, and maybe even much, to be learned from the large number of unillustrated *Physiologus* manuscripts in Europe, at this point of investigation it appears that all observations based on illustrated manuscripts apply equally well to the rest.

This examination of the origin and development of the *Physiologus* begins with a short survey of the hypotheses regarding the background of the Greek *Physiologus* from which the earliest Latin versions were translated. Next the involved question of the various families of Latin manuscripts is reviewed, and the establishment of orderly though inevitably over-simplified groups of versions is presented. Most of the Latin illustrated manuscripts of whose nature I am certain, since they were seen either by means of microfilms or in reality, have been listed in the appropriate place. The total number and kind of extant manuscripts have not, however, been recorded. Closely related to the Latin *Physiologus* are the four principal French bestiaries of the twelfth and thirteenth centuries. These have been viewed with particular reference to their Latin antecedents with both similarities and differences pointed out though not in every case explained because of the numerous puzzles that still tantalizingly exist in this corner of research. A brief outline of the appearance of illustrated manuscripts concludes the first part of this study, which has largely been concerned with the external form of the *Physiologus*. The second part, comprised of the long Chapter Five, the Appendix, and the Plates, presents in alphabetical order résumés of the contents of the Latin and French bestiaries (this term being used to include all versions related to the *Physiologus*) as well as some fifty-six line drawings to show the way in which the often fantastic verbal descriptions were visually portrayed by the mediaeval illustrator. Throughout the work full bibliographical details are given at the first reference to a book or article.

Many people both known and unknown have helped in the course of this research by their information freely shared and, quite as important, by their interest and encouragement over a long period of time. Nor can I forget the very real material aid afforded by two grants from the Smith Research Fund of the University of North Carolina and one from the Committee on Faculty Research of Sweet Briar College for the purchase of microfilm, as well as a Grant-in-Aid from the Southern Fellowships Fund which permitted a summer's work at the Harvard University Library, and lastly a grant from the Research Council of the University Center in Virginia for assistance in publishing this study. For permission to reproduce tracings made from manuscripts belonging to their collections I wish to express my appreciation to the following libraries and persons: the Pierpont Morgan Library, the Trustees of the British Museum, the Bodleian Library, the Warden and Fellows of Merton College, Oxford, the Master and Fellows of Corpus Christi College, Cambridge, the College Council of Sidney Sussex College, Cambridge, the Fitzwilliam Museum, Cambridge, the Dean and Chapter of Canterbury, the Bibliothèque Nationale, the Bibliothèque de l'Arsenal, the Bibliothèque Royale de Belgique, the Royal Library of Denmark, the Burgerbibliothek, Bern, and the Bayerische Staatsbibliothek, Munich. Finally, I wish to acknowledge my gratitude to the teacher whose knowledge and joy in the Middle Ages first opened this period to me, Urban T. Holmes, Jr.

With this book the author pays tribute to Vassar College on the occasion of its Centennial.

FLORENCE McCULLOCH

TABLE OF CONTENTS

MEDIAEVAL LATIN AND
FRENCH BESTIARIES

CHAPTER I

THE GREEK *PHYSIOLOGUS*: ITS CHARACTER
AND ORIGIN

From the early centuries of our era through the Middle Ages the *Physiologus* and its later, expanded form, the bestiary, were among the most popular and important of Christian didactic works. Its importance was perhaps more quantitative than qualitative for its style is impoverished and the mode of thought extremely simple; yet it succeeded in capturing the imagination and interest of men until its understandable disappearance at the time of the Renaissance.[1] For an adequate understanding of the subject of this study — the Latin *Physiologus* and its French translations — the ultimate source of them all, the Greek *Physiologus*, must first be described and its background briefly presented.

The *Physiologus* is a compilation of pseudo-science in which the fantastic descriptions of real and imaginary animals, birds, and even stones were used to illustrate points of Christian dogma and morals. The forty-eight or forty-nine chapters which comprise the Greek *Physiologus* usually follow a set form.[2] A short example will disclose the pattern of a rather typical *Physiologus* chapter. The section on the Pelican begins with a quotation from Psalm 102:6

---

[1] This remark applies more to Latin manuscripts whose period of greatest diffusion was the twelfth and thirteenth centuries —to judge by available manuscripts— and which by the fifteenth century were comparatively rare. On the other hand, one of the most beautifully illuminated Greek manuscripts is Paris, B. N., gr. 834, which was copied in 1585, ascribed to Epiphanius of Cyprus, and edited by Ponce de Leon in Rome in 1587. It is reproduced in Migne, *Patr. Gr.*, XLIII, Cols. 518-534.

[2] All references to the Greek *Physiologus* will be based on the critical edition of Francesco Sbordone entitled *Physiologus* (Milán, 1936). It will hereafter be referred to as Sbordone.

Note: An asterisk (*) indicates that the text is referred to in the Addenda and Corrigenda.

(Vulgate 101:7): "I am like a pelican of the wilderness", and continues with the customary expression, "Physiologus says" that pelicans are very fond of their young. When the children begin to grow, they strike their parents in the face; then they in turn are struck and killed. On the third day their mother (or their father, depending on the manuscript) pierces her side and spills her blood on the dead offspring, thus reviving them. Following this, Isaiah 1:2 is quoted: "I have nourished and brought up children, and they have rebelled against me", and the young pelicans' actions toward their parents are compared to mankind's striking Christ. The allegory continues with the statement that from the side of the crucified Christ came the blood and water of salvation and eternal life. After a further reference to the significance of water and blood, the chapter on the Pelican ends. It might be added at this time that many of the allegories have a far more tenuous relationship to the animal description than the one presented here, as when, for example, the Antelope's horns are equated with the Old and New Testament.

The following list contains the subjects that are treated in the Greek *Physiologus* in the order in which they appear in Sbordone's first version. The English names are those under which each subject appears in Chapter V of the present study.[3]

1.	Lion	14.	Hedgehog
2.	Lizard	15.	Fox
3.	Caladrius	16.	Panther
4.	Pelican	17.	Aspidochelone
5.	Owl	18.	Partridge
6.	Eagle	19.	Vulture
7.	Phoenix	20.	Ant-Lion[4]
8.	Hoopoe	21.	Weasel
9.	Onager	22.	Unicorn
10.	Viper	23.	Beaver
11.	Snake	24.	Hyaena
12.	Ant	25.	Hydrus
13.	Siren and Onocentaur	26.	Ichneumon[5]

[3] There exists a translation into English of the *Physiologus* based on the Greek and other versions, but the edition was limited to only 325 copies. Francis J. Carmody, trans. *Physiologus. The very ancient book of beasts, plants and stones, translated from Greek and other languages* (San Francisco: Book Club of California, 1953). A translation from Greek to German is that of Emil Peters, *Der Physiologus* (Munich, 1921).

[4] In Chapter V the Ant-Lion is included in the discussion on the Ant.

[5] In Chapter V the Ichneumon is included in the discussion on the Hydrus.

27. Crow
28. Turtle-Dove
29. Frog
30. Stag
31. Salamander
32. Diamond
33. Swallow
34. Peridexion Tree
35. Doves
36. Antelope
37. Fire Stones
38. Magnet[6]

39. Sawfish
40. Ibis
41. Goat
42. Diamond
43. Elephant
44. Pearl and Agate
45. Onager and Ape
46. Indian Stone
47. Coot
48. Amos and the Fig Tree
49. Ostrich

Precise knowledge as to the place and date of origin as well as the identity of the author of the Christian *Physiologus* has been lost in time. This has not, however, prevented many hypotheses based on both internal and external evidence from being offered. The reader who wishes more than the brief summary of the early stages of the Greek *Physiologus* which will be presented here can profitably consult the most recent studies in the field from which this résumé will be drawn: the extensive research by Wellmann;[7] a reply to this in the form of a book by Sbordone;[8] and the useful article and bibliography in Pauly-Wissowa by Ben E. Perry.[9] Although any summary must be made at the risk of oversimplifying numerous complex problems, the early history of the *Physiologus*, as it appears today, is as follows. Alexandria in the second and third centuries A. D. was an important center of crosscurrents of learning. Here lived such renowned Christian theologians as Clement and Origen; here flourished traditional lore of the ancient East; and here also was accumulated the wealth of Greek knowledge. A characteristic of scholarship at this period was its preference for allegorical exegesis of the Scriptures, and in like manner nature was interpreted mystically. Creatures of nature were to be explored for what they revealed of the hidden power and wisdom

---

[6] In Chapter V the Magnet is included in the discussion on the Diamond.

[7] Max Wellmann, "Der Physiologos: Eine Religionsgeschichtlich-Naturwissenschaftliche Untersuchung", *Philologus,* Supplementband XXII (1930), 1-116. All references to Wellmann will be to this article.

[8] Francesco Sbordone, *Ricerche sulle fonti e sulla composizione del Physiologus greco* (Naples, 1936). This work will be referred to as Sbordone, *Ricerche.*

[9] Ben E. Perry, "Physiologus", *Real-Encyclopädie der classischen Altertumswissenschaft,* ed. Pauly-Wissowa, XXXIX Halbband (1941), 1074-1129. PW will indicate references to this article.

of God. It was in this atmosphere of the symbolical study of zo-
ology that the *Physiologus* is thought to have arisen. Hommel in
his edition of the Ethiopic *Physiologus*[10] exposed several reasons
(not all equally valid, however)[11] for locating the composition of
the *Physiologus* in Alexandria. Among these are: the inclusion of
animals characteristic of Egypt, like the Crocodile, the Ichneumon,
and the Ibis; the account of the Phoenix, long an Egyptian symbol
of the rising sun and then of the resurrection; the description of
the Onager and the Ape where the Coptic names of the months
are given; and the existence in the *Hieroglyphica* of Horapollo
of similar accounts of birds and beasts.[12]

Wellmann, although he grants that Hellenistic Alexandria was
the "hatching place" for mysticism and dwells at length upon the
predilection of this region for finding in the world of nature elements
of sympathy and antipathy —a trait typical of many tales appearing in the *Physiologus*— proposes a different date and place of
origin. According to him the time of composition is more nearly
the fourth century and the location is in Syria. He believes that
the allegorical system of interpretation in existence there was
stimulated by Origen's teachings after he transferred his school
from Alexandria to Caesarea Stratonis in 230.[13] For well-founded
reasons Perry considers this supposition untenable.[14]

The distant ancestor of the *Physiologus* and related works by
Hermes, Tatian, Timothy of Gaza, and Oppian, was in Wellmann's
opinion a work on the nature of animals, plants, and stones by
the Egyptian writer Bolos of Mendes living in the third or fourth
century B. C. For this book, called Φυσικὰ δυναμερά, Bolos probably took material from Democritus[15] as well as from oral tradition. As a more immediate source of the *Physiologus*, where he
perceives traces of Jewish influence in certain accounts, Wellmann
proposes a book believed to have circulated at an early date and

---

[10] Fritz Hommel, *Die Aethiopische Uebersetzung des Physiologus* (Leipzig, 1877), pp. xv, xvi.

[11] PW, p. 1104.

[12] The evidence presented in the recent translation into English of this
work would seem to cast some doubt on the collection's authenticity as
an Egyptian (then Greek) production. Horapollo, *The Hieroglyphics of
Horapollo*, trans. George Boas (New York, 1950), pp. 29, 30.

[13] Wellmann, pp. 12-14.

[14] PW, p. 1105.

[15] For a study on the confusion existing in antiquity between the literary
ownership of Bolos of Mendes and Democritus, see Wilhelm Kroll, "Bolos
und Demokritos", *Hermes*, LXIX (1934), 228-32.

attributed to the man long considered synonomous with wisdom, Pseudo-Solomon's Φυσιxά.[16] Striking analogies between the *Physiologus* and a work attributed to Hermes Trismegistus, the *Koiraniden* or *Cyranides*,[17] are pointed out by Wellmann, but these are explained by Sbordone[18] as being dependant upon rather than contributing to the *Physiologus*. To name only a few other authors in antiquity whose writings contain details similar to the work that interests us, Herodotus, Aristotle, Plutarch, and Aelian may be mentioned. Many of the *Physiologus* tales were the common property of the ancient world.

The name Physiologus, Φυσιολόγος, "the naturalist", was probably originally used to indicate the pagan author of a work in which were found the characteristics (Φύσεις) of the various animals, and only later was the name applied to the book itself. When the allegories were added by a Christian writer they influenced both the final choice and the description of the contents.[19] This leads naturally to the subject of authorship of the Christian *Physiologus* —a question without any final solution but interesting in the diversity of attributions. In the year 406 Rufinus composed the Seventeenth Homily on Genesis,[20] which is not, as had previously been thought, a translation of Origen.[21] An oft-cited passage on the third nature of the Lion reads: "Nam physiologus de catulo leonis haec scribit". But unfortunately Rufinus gives no specific information as to the identity of Physiologus. Among the Greek and Latin churchmen to whom authorship of the *Physiologus* has been ascribed are: Peter of Alexandria, Epiphanius, Basil, John Chrysostom, Athanasius, Ambrose, and Jerome.[22] An important

---

[16] Wellmann, p. 60.

[17] F. de Mély and Ch.-Em. Ruelle, *Les Lapidaires de l'antiquité et du moyen âge*, Vol. II, *Les Lapidaires grecs* (Paris, 1898); and Louis Delatte, *Textes latins et vieux français relatifs aux Cyranides*, "Bibliothèque de la Faculté de Philosophie et Lettres de l'Université de Liège" (Paris, 1942).

[18] Sbordone, *Ricerche*, pp. 67-75.

[19] This is the opinion of Friedrich Lauchert, who wrote the first general study of the *Physiologus* entitled *Geschichte des Physiologus* (Strassburg, 1889), p. 4. It is supported by Wellmann, p. 5. Other works on the *Physiologus* of a less broad nature than Lauchert's are: Karl Ahrens, *Zur Geschichte des sogenannten Physiologus* (Ploen, 1885), the same author's *Buch der Naturgegenstände* (Kiel, 1892), and Max Goldstaub, "Der Physiologus und seine Weiterbildung", *Philologus*, Supplementband VIII (1899-1901), 339-404.

[20] Migne, *Patr. Gr.*, XII, Col. 257.

[21] PW, p. 1100.

[22] This list is taken from Sbordone, *Ricerche*, p. 172.

notice concerning the existence of an early Latin *Physiologus* and its alleged author is in the first list of books endorsed and prohibited by the church, the famous Decretum Gelasianum of 494.[23] The notice says: "Liber physiologus ab hereticis conscriptus et beati Ambrosii nomine praesignatus, apocryphus". Ambrose was doubtless named here because in describing the fifth and sixth days of Creation in his *Hexaemeron*[24] he used material similar to that of the *Physiologus* if not this work itself.

This introduces another thorny question which has caused debate among scholars —the knowledge and use of the *Physiologus* by the Fathers of the Church. Lauchert would push familiarity with the text back to Justin Martyr in the first half of the second century,[25] while Wellmann would deny direct acquaintanceship to the earliest exegetes, including Origen.[26] In any case, toward the end of the fourth century there are definite parallel passages in the *Physiologus* and the *Hexaemeron* of Ambrose,[27] in the *Hexaemeron* wrongly attributed to Eustathius of Antioch,[28] and in the writings of Epiphanius, to cite only the best known.[29]

The fifth century was the period of translations of the *Physiologus* into other languages of the Near East such as Ethiopic, Syriac, Armenian.[30] For us, however, it is the Latin translations that are of prime interest, and it is to an examination of them that we now turn.

---

[23] Sbordone, *Ricerche*, pp. 168-69, accepts the date and authenticity of the *decretum*, while the results of the research by Dobschütz lead this latter scholar to believe that it was the work of a learned individual in the first half of the sixth century. Ernest von Dobschütz, *Das Decretum Gelasianum de libris recipiendis et non recipiendis* (Leipzig, 1912), p. 348. For a summary of the theological outlook of the author of the *Physiologus*, who has been accused of Gnosticism, see PW, pp. 1101-4.

[24] That he actually used the *Physiologus* seems very probable because of the identity of wording in part of his description of the Partridge (*Hexaemeron*, vi.3.13) and the *Physiologus* chapter. See Wellmann, p. 10, and Francis J. Carmody (ed.), *Physiologus Latinus, Versio B* (Paris, 1939), page 7.

[25] Lauchert, *op. cit.*, p. 68.

[26] Wellmann, pp. 5-10.

[27] Migne, *Patr. Lat.*, XIV, Cols. 134-288.

[28] Migne, *Patr. Gr.*, XVIII, Cols. 707-794.

[29] See Sbordone's chapter "La tradizione parallela dei 'luoghi comuni' presso i Padri della Chiesa", *Ricerche*, pp. 107-138.

[30] See PW, pp. 1116-19, for detailed information on these and other less known translations.

CHAPTER II

THE LATIN *PHYSIOLOGUS* AND THE BESTIARY

It is not definitely known when the first Latin translations of the Greek *Physiologus* appeared, but since Ambrose apparently copied from the *Physiologus* some of the description of the Partridge in his *Hexaemeron* (vi.3.13), which was composed between 386 and 388, this would seem to be evidence of the existence of a Latin version at that early date.[1] According to Lauchert the archetype of the earliest Latin translations of the *Physiologus* existed before 431. As evidence he cites an interpolated list of heretics in the chapter on the Ant where the name of Nestor, whose teachings were condemned at the Third Ecumenical Council at Ephesus in 431, is lacking.[2] However, it has been pointed out that such a conclusion cannot be drawn from this fact since the name of Nestor might have been unintentionally omitted in the original list or left out by a copyist.[3] At present, the oldest extant Latin manuscripts belong to the eighth century.

There are several versions of the Latin *Physiologus*. Most of them contain in general the same material, but there are also important differences which need to be clarified now and for the time when some scholar will attempt to classify all of the Latin manuscripts in existence.[4] Anyone who delves even briefly into the subject of the Latin *Physiologus* realizes the necessity of establishing

---

[1] See Chapter I, n. 24.
[2] Lauchert, *op. cit.*, p. 89.
[3] PW, p. 1120.
[4] Sbordone has made beginnings in this direction, but the results have not been as enlightening as one would hope. See Francesco Sbordone, "La Tradizione manoscritta del *Physiologus* Latino", *Athenaeum*, Nuova Serie XXVII (1949), 246-80.

some order out of the large number of manuscripts with variations small and great in both the order of content and in the details themselves. Families of manuscripts had to be founded and labeled. A little over a hundred years ago Charles Cahier started with three letters, A B and C, to designate the oldest manuscripts known at that time; the rest of his arrangement was rather erratic. Today we have gone through the whole alphabet with some letters carried to the sixth superscription (B[6] in Carmody's *Versio B* was formerly Berlin, Hamilton 77 and is now New York, Morgan 81). To compound confusion, when Sbordone refers to manuscripts M N E, he is indicating the same manuscripts which Carmody letters Y Y[2] Y[3]. There is much to be said for the simplified Four Families of that admirable scholar Montague Rhodes James,[5] although he was only classifying illustrated manuscripts found in England, and his principal divisions are those adopted here. In this study the emphasis —though not exclusively— will be upon major groups of manuscripts. Prominence too will be given to the manuscripts found in France and England and to those versions from which French bestiaries were derived. A final limitation which should be noted at this point is that the writer's main interest has been manuscripts with illustrations, and that the majority of the remarks and conclusions presented here are based on illustrated manuscripts, although they can be applied equally to those with no pictures.

THE OLDEST LATIN VERSIONS.

1. Y Version.

The Latin *Physiologus* was translated more than once from varying Greek versions. The early manuscripts follow the Greek originals closely and betray their kinship by transcribing Greek words. This relationship is apparent in two of the oldest versions, Y and C. Y has been edited by Professor Carmody[6] from three manuscripts: Y—Munich, Lat. 19417, IX cent.; Y[2]—Munich, Lat.

[5] Montague Rhodes James, *The Bestiary* (Edited for the Roxburghe Club. Oxford, 1928).

[6] Francis J. Carmody (ed.), "Physiologus Latinus Versio Y", *University of California Publications in Classical Philology*, XII (1933-44), 95-134.

14388, IX-X cent.; Y³—Bern, Lat. 611, VIII-IX cent.[7] Each of these attributes the work to a different author: Y to Chrysostomus, Y² to St. John of Constantinople, Y³ to an orthodox bishop. The Y version consists of forty-nine chapters (forty-eight by Sbordone's count) closely related in order and content to the eleventh century Codex Mosquensis graecus 432 which Sbordone designates as Π.[8] Among its unusual chapters are Psycomora (Amos and the Fig Tree), Mirmicoleon (Ant-Lion), Ichneumon (Pharoah's rat), and Rana (Frog). There are also many Biblical citations which render the *Vetus Latina* or the pre-Vulgate Bibles.[9] Apart from certain chapters which are common to Y, A, and C, this version evidently disappeared from circulation after the eleventh century and had no influence on other versions.

2. A—Brussels, Bibl. Roy. 10074, f. 140v.-156v. X cent.

This manuscript contains thirty-six chapters, and has been shown by both Sbordone[10] and Carmody[11] to be of composite origin, containing some of the chapters peculiar to Y.[12] Its more immediate interest for us is that although it is not the oldest illustrated Latin *Physiologus*, as it was considered for many years, it does contain some Carolingian drawings noteworthy for their depiction of the text and, in certain cases, the allegory.

---

[7] Sbordone (*Athenaeum*, p. 251) has added three more manuscripts to this version although none contains the full number of chapters. Designated by Sbordone's letters they are: G - Wolfenbüttel, Gudianus 131 (now 4435), f. 159-168, XI cent.; P - Paris, B. N., n. a., lat. 455, f. 3-8v., X cent.; S - Saint Gall 230, f. 510-518, IX cent.

[8] Sbordone, pp. lxviii-lxx.

[9] See Francis J. Carmody, "Quotations in the Latin Physiologus from Latin Bibles earlier than the Vulgate", *University of California Publications in Classical Philology*, XIII (1944-50), 1-8.

[10] Sbordone, p. lxxii.

[11] Carmody, *Versio Y*, p. 98.

[12] The unillustrated B.M., Royal 6 A. xi (XII cent.) contains the same order of chapters as A if the two halves of its contents are reserved. Another manuscript having chapters common to Y and A is Lyon 125, f.189-196v. (XV cent. Unillustrated.)

[13] A is printed in a rather confused manner by Charles Cahier and Arthur Martin in *Mélanges d'archéologie, d'histoire et de littérature* (Paris, 1851-1856), Vols. II-IV. Its illustrations, f. 140v.-147, with the remaining spaces for drawings left blank, appear in Richard Stettiner's *Die illustrierten Prudentiushandschriften* (Berlin, 1905), Plates 172, 177, 178.

3. C—Bern, Burgerbibliothek, lat. 318, f. 7-22v. IX cent.[14]

C represents another translation, quite corrupt, from the Greek with twenty-four of its twenty-six chapters resembling the Ethiopic text.[15] In the printing of this manuscript and in subsequent notices about it, two omissions have occurred —the final two chapters, those on the Cock and the Horse, have not been included.[16] This manuscript is of importance artistically as the first illustrated *Physiologus*, its miniatures showing traces of Alexandrian influence and stylistic affinities with the Utrecht Psalter.[17]

4 Glossary of Ansileubus.

Mention should perhaps be made of the Glossary of Ansileubus (or *Fragments alphabétiques*, as they are sometimes referred to), if only to rectify a misstatement of the past. The Glossary is a collection of twenty-two short *Physiologus* descriptions, arranged alphabetically, and lacking any allegorical interpretation.[18] Lauchert states that they are related to A and B,[19] but it will soon be apparent that A and B do not belong to the same version. The description in the Glossary of the Viper with its unusual composite male and female bust and its crocodile's tail stands closer to Y, A, and C than to B.

---

[14] This manuscript, formerly in the Stadtbibliothek of Bern, is now in the Burgerbibliothek. Its text is printed in Cahier, *Mélanges d'archéologie,* Vols, II-IV.

[15] Lauchert, p. 90.

[16] After listing the contents of the twenty-four chapters, Cahier writes: "Ce qui suit ne se trouve dans aucun autre bestiaire vraiment ancien que je connaisse. C'est 'Galli cantus' et 'Caballus', compilation sans valeur." *Ibid.,* II,95. The text of *Galli cantus* follows verbatim the notice on the Cock in Ambrose's *Hexaemeron* (v.24.88) and that of *Caballus* is copied from Isidore's *Etymologiae* (xii.1.42-48).

[17] Helen Woodruff, "The Physiologus of Bern", *Art Bulletin,* XII (1930), 226-253. See also Chapter IV of this study for the recent conclusions of Professor Dimitri Tselos.

[18] The exerpts are printed in Angelo Mai's *Classici Auctores* (Rome, 1835), Vol. VII, pp. 589-596, and in J. B. Pitra's *Spicilegium Solesmense* (Paris, 1855), Vol. III, pp. 418, 419.

[19] Lauchert, p. 91.

5. B Version.

It is from the complete *Versio B*,[20] which takes its name and part of its substance from manuscript B (Bern, Lat. 233, f. 1-13, VIII-IX cent.), that the main Latin versions in England and France were to develop in the Middle Ages. This particular version with its usual *incipit*: "Etenim Iacob, benedicens filium suum Iudam, ait: Catulus leonis Iudas, filius *de germine meo*, quis suscitabit eum?"[21] and its thirty-six or thirty-seven[22] chapters is perhaps, along with the metrical *Physiologus* of Theobaldus, the most widespread in existence.[23] A typical full title of a manuscript of this kind is: *Incipit Liber Physiologus de Natura Animalium vel avium seu bestiarum* (Bodl., Auct. T.2.23, f. 127. IX cent.). I know of no illustrated manuscripts belonging to this group, but it would be surprising if none at all existed.

The tables on the following page list the contents of the oldest Latin versions or manuscripts of the *Physiologus*.

FIRST FAMILY

Because it seems the clearest way in which to indicate the variations and growth of B, I intend to follow here the large divisions into Four Families created by M. R. James, although the subdivisions within the First Family are my own. These were established for the purpose of grouping similar manuscripts which had not previously been clearly differentiated. Of course there are some hybrid manuscripts which like the Siren do not fit anywhere with ease, but others are Phoenix-like and follow the same pattern.

---

[20] Francis J. Carmody (ed.), *Physiologus Latinus, Versio B* (Paris, 1939). This edition will be referred to as B.

[21] The italicized words are from the pre-Vulgate Bibles.

[22] As the thirty-seventh and final chapter in his *Versio B* Professor Carmody printed an account of the Lacerta (Lizard). This chapter is never found in the B version, which in a complete manuscript ends with Amos propheta (Amos), Adamas (Diamond), and Mermecolion (Pearl). Unfortunately Professor Carmody does not print these sections but sends the reader to the text of B.M., Royal 2 C. xii, reproduced by Mann, to which we shall soon return.

[23] The rhymed version of Theobaldus and the prominent version ascribed to John Chrysostom and known as the *Dicta Chrysostomi*, though they both use B (and sometimes for the latter B-Is) material, will be discussed in the final section of this chapter.

Versio Y   A

Munich, Lat. 19417, s. IX
Munich, Lat. 14388, s. IX-X
Bern Lat. 611, s. VIII-IX.   Brussels 10074, s. X

	Versio Y		A
1.	Leo	1.	Leo
2.	Autolops	2.	Autalops
3.	Piroboli Lapides	3.	Lapides Igniferi
4.	Serra Marina	4.	Serra
5.	Charadrius	5.	Caladrius
6.	Pelicanus	6.	Pellicanus
7.	Nycticorax	7.	Nycticorax
8.	Aquila	8.	Aquila
9.	Phenix	9.	Phenix
10.	Epops	10.	Formica
11.	Onager	11.	Syrene et Onocentauri
12.	Vipera	12.	Vulpis
13.	Serpens	13.	Unicornis
14.	Formica	14.	Castor
15.	Syrena et Onocentaurus	15.	Hyaena
16.	Herinacius	16.	Dorcas
17.	Ibis	17.	Onager
18.	Vulpis	18.	Hydrus
19.	Arbor Peridexion et Columbis.	19.	Simia
20.	Elephas	20.	Perdix
21.	Dorchon (Caprea)	21.	Structocamelon
22.	Achatis Lapis	22.	Salamandra
23.	(Ostrea) Sostoros Lapis et Margarita	23.	Turtur
24.	Adamantinus Lapis	24.	Columba
25.	Onager et Simius	25.	Epopus
26.	(Indicus) Senditicos Lapis	26.	Onager
27.	Herodius id est Fulica	27.	Vipera
28.	Psycomora	28.	Serpens
29.	Panther	29.	Herinatius
30.	Cetus id est Aspisceleon	30.	Arbor Perindex
31.	Perdix	31.	Eliphans
32.	Vultur	32.	Agates
33.	Mirmicoleon	33.	Adamans Lapis
34.	Mustela et Aspis	34.	Lapis Sindicus
35.	Monoceras	35.	Herodius
36.	Castor	36.	Panthera
37.	Hyena hoc est Belua		
38.	(Hydrus) Niluus		
39.	(Ichneumon) Echinemon		
40.	Cornicola		
41.	Turtur		
42.	Hyrundo		
43.	Cervus		
44.	Rana		
45.	Saura id est Salamandra		
46.	Magnis Lapis		
47.	Adamantinus Lapis		
48.	Columbae		
49.	Saura Eliace hoc est Anguilla Solis		

Versio B   C

Bern 233, s. VIII-IX	Bern 318, s. IX
1. Leo	1. Leo
2. Autolops	2. Aesaura
3. Lapides Igniferi	3. Calatrius
4. Serra	4. Pelicanus
5. Caladrius	5. Nocticorax
6. Pelicanus	6. Aquila
7. Nycticorax	7. Yppopus
8. Aquila	8. Vipera
9. Phoenix	9. Serpens
10. Upupa	10. Formica
11. Formica	11. Serena et Honocentaurus
12. Sirenae et Onocentauri	12. Yricius
13. Herinacius	13. Vulpis
14. Ibis	14. Panthera
15. Vulpis	15. Aspidohelunes
16. Unicornis	16. Unicornium
17. Castor	17. Cervus
18. Hyaena	18. Salamandra
19. Hydrus	19. Peredixion
20. Caprea	20. Antelups
21. Onager et Simia	21. Serra
22. Fulica	22. Elephas
23. Panthera	23. Lapis Acatus et Margarita
24. Aspis Chelone	24. Lapis Indicus
25. Perdix	25. Galli Cantus
26. Mustela et Aspis	26. Caballus
27. Asida	
28. Turtur	
29. Cervus	
30. Salamandra	
31. Columbae	
32. Peredixion	
33. Elephas	
34. Amos	
35. Adamas	
36. Margarita (Mermecolion)	

The *Etymologiae* of Isidore of Seville was the work which effected the first change in the content of the *Physiologus*.[24] This collection of twenty books, which was composed by the bishop of Seville († 636), exerted an influence far into the Middle Ages. It is a type of dictionary where the words are arranged not alphabetically but by subjects. Book XII, *De animalibus*, was the source of the new additions to the *Physiologus*. The form taken in each of the several divisions of this book is first to present an etymology of the animal's name —it is almost always of fantastic composition (as, for example, "Formica, eo quod ferat micas farris", 'The ant is thus called because it carries bits of wheat')— followed by a short description[25] also of an imaginative nature. Many of Isidore's passages already resemble the *Physiologus*, while Pliny's *Historia naturalis* seems the ultimate source for others.[26] The intermediary through which some of this information descended was the well known *Collectanea rerum memorabilium* or *Polyhistor* of Solinus, who repeated much of Pliny's material on animals.[27] These are only a few of the strands that were pulled together by Isidore to reappear in the growing *Physiologus*.

1. B-Is Version

In the hope that more confusion will not be caused by the creation of yet another set of letters, the designation B-Is is proposed to describe those manuscripts which follow the order and

---

[24] W. M. Lindsay (ed.), *Etymologiarum sive originum libri XX*. 2 vols. (Oxford, 1911). For a translation into English of parts of this work, see Ernest Brehaut, "An Encyclopedist of the Dark Ages: Isidore of Seville", *Columbia University Studies in History, Economics and Public Law*, XLVIII (1912), 7-274.

[25] Many of Isidore's etymologies are taken from Varro's *De lingua latina*.

[26] Pliny, *Naturalis historiae libri XXXVII*, ed. Carolus Mayhoff (5 vols., Leipzig, 1892-1909). An English translation with occasionally helpful footnotes is that made by John Bostock and H. T. Riley (6 vols., Bohn Library, London, 1855). A more recent translation is that in the Loeb Classical Library by H. Rackham and W. H. S. Jones.

[27] Solinus, *Collectanea rerum memorabilium*, ed. Th. Mommsen (Berlin, 1864). See also Lynn Thorndike, *A History of Magic and Experimental Science* (New York, 1929), Vol. I, pp. 326-28. A facsimile reproduction has been published of Arthur Golding's 1587 translation of Solinus into English: *The Excellent and Pleasant Worke of Caius Julius Solinus* (Gainesville, Florida: Scholars' Facsimiles and Reprints, 1955).

content of B but which, except for seven chapters,[28] contain additions from Isidore. Sometimes the added passages are not indicated as such, but many manuscripts, after a description and allegory identical with B, have rubrics reading *Ethimologia ysidori* (B. M., Laud Misc. 247) or a blank space where these words or *Ethimologia* alone were intended. When the additions were first made is not known, although the oldest manuscript that has been seen which shows this change is Vatican, Palat. lat. 1074, f. 1-22, X cent.[29]

The only printed example of this essential text was prepared by an early student of the Latin *Physiologus* tradition, Max Friedrich Mann, whose object at the time was to show the almost total dependence of Guillaume le Clerc's *Bestiaire* on a version of this type. The manuscript used by Mann was the unillustrated, early thirteenth century B. M. Royal 2 C. xii, f. 133-145v.[30] As will be indicated later, the bestiaries of Philippe de Thaon and Pierre de Beauvais were also largely based on the B-Is version.

In the following list, illustrated manuscripts of the B-Is version are grouped according to iconographic resemblances, although these are rather faint in some cases. Probably more of such illustrated manuscripts exist.

> Bodl., Laud Misc. 247, f. 139v.-166v. Early XII cent.
>
> Bodl. 602, f. 1-36. Late XII cent. (*Aviarium*, f. 36-65).
>
> Bodl., Douce 167, f. 1-12. Early XIII cent. (Adds chapter on Lupus after Mermecolion).

---

[28] The chapters containing no additions from Isidore are: Antalops, Lapides Igniferi, Serra, Caladrius, Peridexion, Amos, and Mermecolion.

[29] Willy Lüdtke, "Zum armenischen und lateinischen Physiologus", *Huschardzan Festschrift* (Vienna, 1911), p. 220. A point of particular interest in this manuscript is that it shows how the large Second Family bestiary could first have been composed. At the end of the entire *Physiologus* with its fixed passages from Isidore, the scribe has continued (f. 21v.) *De etimologiarum libro,* and has apparently aimlessly copied and somewhat elaborated on Isidore's descriptions of the Psitacus, Ercine [sic], and Coturnix — birds totally unrelated to the old *Physiologus* contents. All that was then needed was a more systematic borrowing from Isidore, which would result in the highly organized Second Family bestiary to be discussed later.

[30] Max Friedrich Mann, "Der Bestiaire Divin des Guillaume le Clerc", *Französische Studien*, VI Band, 2 Heft (1888), 37-73.

London, Sion College L $\dfrac{40.2}{\text{L }28}$, f. 73-116. XIII cent.
(*Aviarium*, f. 1-54).

(Dyson) Perrins 26, f. 67-102v. XIII cent. (*Aviarium*, f. 1-45).[31]

The next two manuscripts show the addition of new articles on Lupus, Canis, Ibex, Noctua (with Nicticorax), and Cocodrillus (separated from Hydrus). Their beginning chapters also differ in following an order which resembles that of the following group, Book II of Pseudo-Hugo.

> B. M., Stowe 1067, f. 1-15v. Early XII cent. (Illustrations and script resemble Bodl., Laud Misc. 247.)
>
> Camb., Corpus Christi Coll. 22, f. 162-169, XII cent.

2. H—*Book II of Pseudo-Hugo of Saint Victor and the* Aviarium.

The *Physiologus* printed in Migne as Book II of the *De bestiis et aliis rebus* incorrectly attributed to Hugo of Saint Victor is a composition of a rather special nature.[32] Although much of the material closely copies the B-Is version, with the Isidorean additions

---

[31] This manuscript of French provenance was sold at Sotheby's in London on December 9, 1958, for £36,500. The purchaser was Mr. H. P. Kraus of New York.

[32] One probably wonders how the *Physiologus* or bestiary, a work now regarded as most certainly anonymous, came to be attributed to Hugo, the noted twelfth century theologian and head of the famous monastery of Saint Victor in Paris. An uncritical edition of Hugo's writings containing both genuine and spurious works was published in 1648 in Rouen and was republished in 1854 by Migne, who kept in Volume CLXXVII the four books comprising the collection entitled *De bestiis et aliis rebus*. In the introduction to Hugo's works (CLXXV, cxviii) it is stated that the Benedictines in naming authors of the different sections of the *De bestiis* assigned Book I to Hugo of Folieto, Book II to Henry of Ghent, and Books III and IV to Guillelmus Peraldus (the latter is no more than a list of short items based on the previous books and will not be considered here). Apparently the second of these statements is based on a misreading since the *Histoire littéraire de la France* under Hugues de Fouilloi says, concerning Book II: "on l'attribue communément à Alain de Lille, qui, selon Henri de Gand, a composé un livre de la nature de quelques animaux, qu'on croit être celui-là." (*HLF*, XIII, 498). However, the French scholar B. Hauréau examined Alain de Lille's book on animals, *Distinctiones*, and found that it

often preceding rather than following the traditional account and with Isidore and Solinus both named on occasion, the order is decidedly different. Book II, which for the sake of convenience will be called H, contains some of the added chapters that we have seen before, as well as others, Ibex, Canis, Lupus, Vipera, Lacerta, and Draco separated fron Panthera, but it is very different from B-Is in only describing two birds.

To explain this peculiarity we must now digress and enter the realm of the *Aviarium*. The first book of the *De bestiis* is a compilation limited to birds and is thus known as the *Aviarium* or, because the first eleven chapters treat Doves and their characteristics, the *Columba deargentata*, the "Silvered Dove" (Ps. 67:14, Vulgate).[33] After a prologue in the form of a letter to a certain Raynerus, a convert, that sometimes precedes the text and begins "Desiderii tui karissime...", the text itself starts: "Si dormiatis inter medios cleros...". The moralizations in the *Aviarum* are greatly expanded, and Rabanus Maurus and the *Moralia* of Gregory are quoted. Although the *Aviarium* exists separately as well as before or after at least two versions of the *Physiologus*, it is its close relationship with H that concerns us here. Professor Carmody glimpsed what has proved to be the explanation when he concluded that the compiler of Book II had Book I before him and consciously avoided duplicating the birds in Book II that he had treated in I.

---

bore no resemblance to a bestiary. The bestiary which Hauréau used for this comparison is in the Bibliothèque Mazarine, MS. 742, an unillustrated, thirteenth century manuscript from the library of St. Victor (see *Journal des Savants*, 1887, p. 33), and although it was not realized at the time, this is a typical Second Family bestiary. The only explanation that I can find for originally including manuscripts of two different versions of the bestiary in the works of Hugo is the existence of one manuscript, though there are probably others, mentioned in Manitius with a rubric in a contemporary hand reading: "Incipit bestiarius hugonis de sancto victore" (apparently this manuscript, Dresden A 198, f. 47-81, XIII cent., is a bestiary and *Aviarium*. See Max Manitius, *Geschichte der Lateinischen Literatur des Mittelalters*, III, 227). Thus an ancient error was perpetuated in Migne's nineteenth century edition.[**]

[33] See S. Ives and H. Lehmann-Haupt, *An English 13th Century Bestiary* (New York: H. P. Kraus, 1942). The manuscript described in this interesting study was at that time called the Kraus Bestiary. It is now in the possession of Mr. and Mrs. Philip Hofer of Cambridge, Massachusetts, and will be referred to as the Hofer Bestiary. Some of the manuscripts attributed to Hugo of Folieto are listed in Manitius, *op. cit.*, III, 227-28, and others in Thorndike, *op. cit.*, II, 17-18. A study of the various manuscripts of the *Aviarium* has been undertaken in Paris by M. Yves Lefèvre.

The Pelican and Caladrius are, however, found twice.[34] M. R. James was even closer when, in describing a manuscript of this group, he said: "At least 4 leaves are gone, and, it is consequently difficult to see exactly where the treatise addressed to Rainer ends, and where the Bestiary begins — if indeed the whole be not one work."[35] And, to judge by two complete manuscripts, the whole is assuredly one work.[36] In both B. N., lat. 14429 and Valenciennes 101 there is no break in the text between the *Aviarium* and the beginning of the *Physiologus* other than a new paragraph. The former then continues: "Leo ex greco vocabulo inflexum..." and the latter: "Incipit liber de naturis bestiarum qui phisiologus appellatur. bestiarum autem vocabulum..."

The illustrated manuscripts known to exist are the following:

> Camb., Sidney Sussex Coll. 100, f. 26-43. XIII cent. (*Aviarium*, f. 1-26).
>
> Valenciennes 101, f. 189-201. XIII cent. (*Aviarium*, f. 171-189).
>
> B. N., lat. 14429, f. 109v.-118. XIII cent. (*Aviarium*, f. 96-109).
>
> Chalon-sur-Saône 14, f. 55-89. XIII cent.*

The table of contents, with some simplification in the titles for Book I, is taken from Migne.

---

[34] Francis J. Carmody, "De Bestiis et Aliis Rebus and the Latin Physiologus", *Speculum*, XIII (1938), 155.

[35] Montague R. James, *A Descriptive Catalogue of Manuscripts in the Library of Sidney Sussex College, Cambridge* (Cambridge, 1895), p. 116.

[36] Whether Book II once existed as a *Physiologus* version independent of the *Aviarium* as Ives and Lehmann-Haupt suggest (*op. cit.*, p. 16), I do not know. However, evidence that H was probably originally based on a B version manuscript is the unexpected appearance in the moralization of the Antelope (2) of two sentences on the Fire Stones (Lapides Igniferi), an intrusion quite unrelated to anything that precedes. It will be recalled that the inevitable third chapter, following the Antelope, in a manuscript of the B version is that on the Fire Stones. The same anomaly occurs in the chapter on the Antelope (9) in the *Dicta Chrysostomi* which will be discussed later.

Book I of *De bestiis et aliis rebus*:   Book II of *De bestiis et aliis rebus*:

1.-11. De columba
    12. De aquilone et austro ventis
13.-22. De accipitre
23.-25. De tuture
26.-32. De libano et cedro et pas-
         seribis
    33. De pelicano
    34. De nycticorace
    35. De corvo
    36. De gallo
    37. De struthione
    38. De vulture
    39. De grue
    40. De milvo
    41. De hirundine
    42. De ciconiis
    43. De merula
    44. De bubone
    45. De graculo
    46. De ansere
    47. De ardea
    48. De caladrio
    49. De phoenice
    50. De perdice
    51. De coturnice
    52. De upupa
    53. De olore
    54. De classe Salomonis
    55. De pavone
    56. De aquila
    57. De ibe seu ibide
    58. De fulica

 1. De leone
 2. De antula
 3. De onocentauro
 4. De herinaceo seu hericio
 5. De vulpe
 6. De monocerote sive unicorni
 7. De hydro et hydra
 8. De crocodili natura
 9. De castoris natura
10. De hyaena
11. De onagro
12. De simiis
13. De capri natura
14. De cervorum natura
15. De ibice
16. De stellione et salamandra
17. De canibus
18. De mustela et aspide
19. De lapidis igniferis
20. De luporum natura
21. De viperae natura
22. De serra bellua marina
23. De pantherae natura
24. De dracone
25. De elephantis natura
26. De elephantis natura iterum
27. De pelicani natura
28. De lacerto, stellione et lacerta
29. De formicae natura
30. De aspidis natura
31. De charadrio seu charadro
32. De sirenarum seu sirenum na-
     tura
33. De onocentauro rursus
34. De adamantis virtute
35. De concha seu concha marga-
     ritifera
36. De aspidochelone

3. *Transitional manuscripts.*

Transitional is the only name that adequately but dully describes these manuscripts which in their appearance are of fine and imaginative execution. From the First Family they keep the first twenty-four to forty chapters, following the order and text of either B-Is or H. Then they continue with sections taken from Isidore's *Etymologia,* starting with his second chapter on Beasts and describing in order Tigris, Pardus, Linx, Grifes... Morgan 81 and its sister manuscripts, grouped together below, have sections at the beginning from Isidore which occur nowhere else. The entire contents are classified into Beasts, Birds, Fish, and so on. This classification will be more fully discussed when the characteristics of the Second Family are enumerated.

The following are illustrated Transitional manuscripts:

> Camb., Trinity Coll. R.14.9 (884), f. 89-106v. XIII cent. (Beginning follows in general order of B to Amos, then begins Tigris, etc.)

> B. M., Royal 2 B. vii, f. 85-130v. Early XIV cent.[37] (Marginal drawings. No *Physiologus* text. Follows with a few exceptions order of B-Is through Elephant and Mandrake, then adds nine birds and animals based, in part my opinion, on a text by Pierre de Beauvais.)

> New York, Morgan 81, f. 1-89. Late XII cent. (Formerly Berlin, Hamilton 77. Beginning remotely follows order of H before continuing with Tigris.)

> Leningrad, Qu.V.1, f. 1-98. Late XII cent.[38]

> B. M., Royal 12 C. xix, f. 1-94. Late XII cent.

> Alnwick Bestiary, f. 1-73. Mid. XIII cent.[38a]

> Munich, gall. 16 (Queen Isabella's Psalter), XIV cent.*

SECOND FAMILY - THE BESTIARY

The nature of the old *Physiologus* changes sometime during the twelfth century —not beyond recognition because the venerable B-Is chapters can always be perceived within the mass of added material, but the transformation is still very great. Though the analogy should not be pushed too far, the change might be compared to the one taking place in architecture at approximately the same time— from the simpler Romanesque to the more highly developed Gothic. The number of chapters in what is now properly called the bestiary is far more than doubled, with most of the additions coming from

---

[37] A facsimile reproduction of this manuscript exists: *Queen Mary's Psalter*. Introduction by Sir George Warner (London: The British Museum, 1912).

[38] For a description of this manuscript see Alexandra Konstantinova, "Ein englisches Bestiar des zwölften Jahrhunderts", *Kunstwissenschaftliche Studien*, IV (1929).

[38a] This ms., almost identical with 12 C. xix, has been reproduced in facsimile. See *Bibliography*, Millar.

Isidore. Among the characteristics of this family that James notes are: the classification of the entire contents following the divisions of Isidore's Book XII, the inclusion of many chapters with no moral or spiritual exposition, the addition of material from Solinus, long extracts from Ambrose's *Hexaemeron*, in some copies reflections from Rabanus Maurus and from Peter of Cornwall's *Pantheologus* (there is a certain amount of variation among the copies), and the attachment "without rubric, rhyme, or reason" of a sermon beginning *Quocienscumque peccator* to the article on Canis.[39] James declares that the twenty illustrated manuscripts which he includes in the Second Family are all of English origin. I am unable to say whether all four of the examples that I have noted in France and Belgium are English or not, but for evidence of certain similarities that were found, see Chapter IV, note 17.

The bestiary printed as Book III of the *De bestiis* is not reliable as a complete text because, as Professor Carmody has pointed out, the manuscript used for Books I and II to which the editor refers for all of the chapters repeated in Book III was a different one from that partially printed in Book III.[40] This means that for the text in its entirety one must at present supplement Migne's Book III with James's reproduction of Cambridge, University Library Ii.4.26.[41]

Before listing the contents which usually appear in a Second Family manuscript, a few general remarks about the form of this type of bestiary are needed. The *incipit* is one or both of two phrases: "Leo fortissimus bestiarum ad nullius pavebit occursum" (Proverbs 30:30 as cited in the *De Universo* [VIII.1] of Rabanus Maurus), or "Bestiarum vocabulum proprie convenit pardis..." (Isidore, xii.2.1). Apart from a few pictures in the section on Fish (often only the Aspidochelone), the illustrations usually stop at the Snake, with an occasional portrayal at the very end of the Fire Stones. The long sections on Trees and the Ages of Man, both from Isidore and forming the concluding divisions of many bestiaries, will not be included in this study.[42]

---

[39] James, *Bestiary*, p. 14.

[40] Carmody, "De Bestiis...", p. 158.

[41] This manuscript has been translated into English in a lively but not always wholly accurate manner by T. H. White in a book entitled *The Book of Beasts* (London, 1954). See the present writer's review in *Speculum*, XXX (1955), 694-96.

[42] Since certain limits in content had to be made in this study, many of the topics described in a sentence or so in Isidore and consequently in the Second Family bestiaries and thereafter are omitted in Chapter V

An attempt has been made to group the following illustrated manuscripts of the Second Family according to iconographic resemblances, but since for a few of them written descriptions alone had to be relied upon, these groupings should not be judged as definitive.

> B.M., Add. 11283, f. 1-41. Early XII cent. (The earliest Second Family manuscript known to me.)[43]
>
> Brussels, Bibl. Roy. 8340, f. 183-215. XIV cent.
>
> Aberdeen Univ. 24, f. 1-103. Late XII Cent.
>
> Bodl., Ashmole 1511, f. 1-104. Late XII cent.[44]
>
> Bodl., Douce 151, f. 1-90. XIV cent. (Poor copy of above.)
>
> Oxford, Univ. Lib. 120, f. 1-70. XIII-XIV cent.
>
> B.M., Harl. 4751, f. 1-74v. Late XII cent.
>
> Bodl. 764, f. 1-137. Late XII cent.
>
> Camb., Univ. Lib. Ii.4.26, f. 1-74. XII cent.
>
> Oxford, St. John's Coll. 61, f. 1-103. XIII cent.[45]
>
> B.M., Harl. 3244, f. 36-71v. Early XIII cent.
>
> Camb., Gonv. and Caius Coll. 109, f. 110-133. (Only 5 illustrations.) XIII cent.
>
> B.N., lat. 3630, f. 75-96. XIV cent.

---

of this paper. Such is the fate of most of Isidore's Chapters III, V, VI, and VIII of Book XII. Other articles are not mentioned when they exist in only one or two of the Latin bestiaries which were examined since the typical text was the object of research rather than the exceptional. This accounts for the omission in Chapter V of the following subjects as found in the unusually full manuscript Bodl. 764: Tragelaphus, Lepus, Vitulus, Taxus, Glires, Aurifrigius, and Martineta. The account of these last two birds, preceded in the manuscript by Bernaca (Barnacle Goose), would appear to be drawn from the *Topographica Hibernica*, Dist. I, Chapters XVI and XVII of Giraldus Cambrensis, since the three birds appear in that order with a similar text in the Welshman's work. Many of the additions such as Dama, Cuniculus, and Cirogrillus that appear in Third Family manuscripts have not been treated here either.

[43] The latest Second Family bestiary which I have found is Nîmes 82 (13777), an unillustrated manuscript dating from the beginning of the sixteenth century (see *Catalogue général des manuscrits des bibliothèques publiques des départements.* Vol. VII (1885), p. 577).

[44] The Aberdeen, Ashmole, Leningrad and Alnwick manuscripts begin with excellent scenes of the Creation of the World while those found in Gonville and Caius 372 are much more modest in conception and execution. These manuscripts, with the exception of those in Leningrad and Alnwick, also largely follow the order of the *Aviarium* in the section devoted to Birds.

[45] Iconographically these two manuscripts show some similarity to Morgan 81 and Leningrad Qu.V.1.

The following manuscripts are unrelated iconographically:

> B.M., Royal 12 F. xiii, f. 1-141. XII-XIII cent. (Illustrations end at f. 50.)
> B.M., Sloane 3544, f. 1-44. XIII cent.
> Camb., Gonv. and Caius Coll. 384, f. 167-199. XIII cent.
> Camb., Gonv. and Caius Coll. 372, f. 1-64. XIII cent.
> B.N., lat. 11207, f. 1-40. XIII cent.
> Bodl. 533, f. 1-29v. XIII cent.
> Oxford, St. John's Coll. 178, f. 157-220. Late XIII cent.
> Douai, Bibl. Mun. 711, f. 1-60v. Late XIII cent.
> Bodl., Douce 88 A, f. 5-29. Late XIII cent.
> Canterbury, Cath. Lib. Lit. D 10. XIII-XIV cent.
> Camb., Corpus Christi Coll. 53, f. 189-210. Early XIV cent.[46]
> New York, Morgan 890, f. 1-18. XIV cent.[47]
> Copenhagen, Gl. Kgl. 1633 4°, f. 1-76v. XIV cent.

The following table of contents illustrating a typical Second Family manuscript is taken from the total list of chapters contained in Book III of *De bestiis et aliis rebus* ascribed to Pseudo-Hugo of Saint Victor. The principal divisions of the work are italicized.

De bestiis.

1. De leone	16. De capro
2. De tigride	17. De monocerote sive unicorni
3. De pardo et leopardo	18. De urso
4. De panthera	19. De leucrocuta
5. De antalope seu antula	20. De crocodilo
6. De unicorni	21. De manticora
7. De lynce	22. De tharando
8. De gryphe	23. De vulpe
9. De elephante	24. De eale animali
10. De castore	25. De lupo
11. De ibice	26. De cane
12. De hiena	27. *De animalium in genere.* "Omnibus animantibus Adam..."
13. De bonasa	
14. De simiis	28. De ove
15. De cervis	29. De vervece

[46] This manuscript has been reproduced with the following title: *A Peterborough Psalter and Bestiary of the Fourteenth Century.* Description by Montague Rhodes James (Oxford: Roxburghe Club, 1921).

[47] Formerly belonging to Sir Sidney Cockerell's collection, this manuscript was given to the Morgan Library in 1958.

30. De agno
31. De hirco et haedo
32. De apro
33. De juvenco et tauro
34. De bove et uro
35. De cameli natura
36. De dromedario
37. De asino et asello
38. De onagro
39. De equo
40. De cato seu musione
41. De mure et sorice
42. De mustela
43. De talpa
44. De formica
45. De hericio seu herinaceo
46. *De avibus in genere.* "Avium unum quidem..."
47. De aquila
48. De vulture
49. De gruibus
50. De psittaco
51. De caladrio
52. De ciconiis
53. De holore
54. De ibide seu ibi
55. De assida seu struthione
56. De fulica
57. De halcyne
58. De phoenice
59. De cinnamulgo
60. De herciniis
61. De epope ... upupa
62. De pelicano
63. De noctua seu nycticorace
64. De syrenis
65. De perdice
66. De pica et pico
67. De accipitre
68. De luscinia
69. De vespertilione
70. De cornice et corvo
71. De columba
72. De turture
73. De hirundine
74. De coturnice seu qualea
75. De pavone
76. De upupa
77. De gallo
78. De anate
79. De ovis et ex eis natis
80. De apibus
81. De arbore quadam in India
82. *De serpentum generibus.* "Anguis omnium serpentum..."
83. De dracone
84. De basilisco et sibilo
85. De vipera
86. De aspide
87. De ceraste
88. De scitale
89. De amphysibaena
90. De boa serpente
91. De jaculo
92. De sirenis serpentibus
93. De sepe serpente
94. De dypsade serpente
95. De lacerto et batracha
96. De salamandra
97. De saura
98. De stellione iterum et aliis serpentibus
99. De serpentum varia natura
100. *De vermibus.*
101. *De piscium diversorum naturis... et concharum.*
102. *De arboribus.*
103. De margaritarum inventione
104. De lapidibus igniferis
105. De duodecim lapidibus pretiosis
106. *De natura hominis.*
107. De hominis membris ac partibus
108. *De aetatibus hominis.*

THIRD AND FOURTH FAMILIES

These even larger bestiaries, all of the thirteenth century, will be treated briefly here since they are less common and since they have nothing that was carried over to the French translations. A complete manuscript begins with the words "Cum voluntas conditoris", and the first section repeats Isidore's account of the Fabulous Nations who inhabit the remote parts of the earth (Isidore xi.3.1-39),[48] followed by the discourse on animals beginning "Omnibus animantibus...". Next come extracts from the *Megacosmus* or *De mundi universitate* by Bernardus Silvestris, who is called Bernardus francus here. The bestiary proper begins with the Domestic Animals, Bos, Bubalus, Vacca..., and continues with the Wild Beasts, Leo, Pardus, Linx, Panthera.... The following sections are on Fish, Snakes, and Insects. Another section from Isidore on mythological monsters like Cerberus and the Chimaera is followed by Lapides Igniferi. The composition of the last part of the book varies in the few extant copies, but apparently the following items included in Westminster 22 belong to it: the Wheel of Fortune, a portion of Seneca's *De remediis fortuitorum*, the Seven Wonders of the World, and a passage from the *Policraticus* of John of Salisbury.

Of the five known manuscripts (James did not list Bodl. e Museo 136 which might have been written in the Netherlands and is ascribed to Hugo of Folieto) the illustrations in all except the Westminster manuscript resemble one another markedly.

> Camb., Fitzwilliam Museum 254, f. 1-48. Early XIII cent.
> Camb., Univ. Lib. Kk.4.25, f. 48-86. XIII cent.
> Bodl., Douce 88 E, f. 68-116v. Late XIII cent.
> Bodl. e Museo 136, f. 1-47. XIII cent.
> Westminster Abbey 22, 1-54. XIII cent.

The one manuscript belonging to the Fourth Family as described by James is Camb., Univ. Lib. Gg.6.5, f. 1-100 (XV cent.) which is based as much upon Bartholomeus Anglicus' *De proprietatibus rerum*, named in the text as a source, as upon Isidore. The *sensus moralis* is often omitted *propter prolixitatem*. The unfinished text ends with a chapter on Trees.[49]

---

[48] For illustrations and a discussion of these prodigies as they appear on the Icelandic *Physiologus*, see Halldor Hermannson, "The Icelandic Physiologus", *Icelandica*, XXVII (1938),12-14. Additional information may be found in George C. Druce, "Some Abnormal and Composite Human Forms in English Church Architecture", *Archaeological Journal*, LXXII (1915), 135-86.

[49] As noted in Chapter IV of this study, the large, careless pictures in this manuscript look like poor relations of those found in Copenhagen, Gl. Kgl. 1633 4° (a Second Family manuscript).

Thus conclude the successive changes undergone by one version of the *Physiologus* from the bare, incomplete B manuscript of the eighth or ninth century to the inflated and eclectic Fourth Family manuscript of the fifteenth. There were also two other important versions current during the Middle Ages which deserve attention now.[50]

OTHER PRINCIPAL VERSIONS

1. TH - Metrical Physiologus *of Theobaldus.*

Numerous are the manuscripts with the title *De naturis (proprietatibus) animalium,* or *Bestiarius* (sometimes *Physiologus) seu liber de naturis XII animalium* with or without the name of Theobaldus attached. This poem of about three hundred lines in mixed meters is usually attributed to the Theobaldus who was abbot of Monte Cassino from 1022 to 1035 and under whom studies of natural science and medicine were pursued in the monastery.[51]

---

[50] Sbordone, *Tradizione manoscritta,* pp. 277-79, lists some Vatican manuscripts with uncommon additions which will not be discussed here. Nor will the interesting, illustrated Berlin, Hamilton 390 described by A. Tobler, "Lateinische Beispielsammlung mit Bildern", *Zeitschrift für Romanische Philologie,* XII (1888), 57-88. In reality only twelve of the forty-four chapters in this manuscript are devoted to animals belonging strictly to the *Physiologus.* Of this number, some of the moralizations are illustrated. For the relationship of this manuscript with numerous fourteenth and fifteenth century bestiaries in Italian, see Kenneth McKenzie, "Unpublished Manuscripts of Italian Bestiaries", *PMLA,* XX (1905), 383.

[51] Manitius, III, 731. A list of manuscripts attributed to Theobaldus is found on p. 734. Sbordone, *Tradizione manoscritta,* pp. 273-75, adds more manuscripts to this list, although B.N., lat. 10448 should be removed. The Italian scholar (p. 273) estimates that there are at least one hundred manuscripts of this version dating from the twelfth to the fifteenth century scattered in various libraries throughout Europe. The following manuscripts of Theobaldus now at the Library of the University of Uppsala were not mentioned by Sbordone: C 181, f. 321-325, XV cent.; C 218, f. 2-18, XV cent.; C 226, f. 14v.-15, XV cent.; C 237, f. 112-132, XIV cent. The Latin text of Theobaldus' poem is printed in the works of Hildebert of Tours or Le Mans (Migne, *Patr. Lat.,* CLXXI, Cols. 1217-1224), and in Richard Morris, *An Old English Miscellany* (Early English Text Society. London, 1872), pp. 201-9. An English translation has been made by Alan W. Rendell, *Physiologus a Metrical Bestiary of Twelve Chapters by Bishop Theobald* (London, 1928).

After explaining the meaning of the word 'Physiologus' thus: "...et dicitur a phisis grece, quod est natura latine, et olon, quod est totum, et logos quod est sermo, quia sermo totus de natura", the poem begins: "Tres leo naturas et tres habet inde figuras". The animals described are:

1.	De leone	7.	De aranea
2.	De aquila	8.	De balena
3.	De serpente	9.	De sirenis et de onocentauris
4.	De formica	10.	De elephante
5.	De vulpe	11.	De turture
6.	De cervo	12.	De panthera

2. DC - Dicta Chrysostomi.

Differing in some details[52] and in order from B is a work attributed to the early fifth century Patriarch of Constantinople, John Chysostom, and circulating under the title *Dicta Chrysostomi* (from the *incipit* which usually reads: "Incipiunt dicta Johannis Crisostomi de naturis bestiarum").[53] The principal characteristic of this work is the division of the contents into Animals beginning with Leo, and Birds starting with Aquila. Stones are omitted entirely. Of special interest for this study is the fact that the place of origin of the *Dicta Chrysostomi* is regarded as France *ca.* 1000.[54] Because Lauchert omitted any reference to it in his original study on the *Physiologus,* some scholars have failed to note that there exists a French rhymed bestiary largely resembling the *Dicta.* Its translator, Gervaise, ascribes his original to John Chrysostom. Two French

---

[52] TH and DC are unique in their notice about the Eagle's beak growing long with age; they also differ somewhat in their chapter on the Stag.

[53] The most recent research on the *Dicta Chrysostomi* is that of Hermann Menhardt, "Der Millstätter Physiologus und seine Verwandten", *Kärntner Museumsschriften,* XIV (1956), from which some details are taken. One text of the *Dicta Chrysostomi* has been printed by Gustav Heider, "Physiologus nach einer Handschrift des XI. Jahrhunderts", *Archiv für Kunde österreichischer Geschichts-Quellen,* Dritter Jahrgang, zweiter Band, 1850, pp. 541-82. As his text Heider used what he termed the eleventh century Codex Gottwicensis 101. This manuscript is now in the Pierpont Morgan Library, MS. 832, and is assigned to the twelfth century. Heider reproduced a few more than half of the illustrations in this interesting manuscript. A critical edition is that of Friedrich Wilhelm, *Münchener Texte,* Heft 8 B (Kommentar), 1916, pp. 15-44.

[54] Wilhelm, *op. cit.,* p. 16 and Menhardt, *op. cit.,* p. 76.

manuscripts of the short bestiary of Pierre de Beauvais also name the Greek churchman as their source.

The expected number of chapters included in the DC is twenty-seven, although many copies have a variety of additions.[55] One of the most interesting manuscripts of DC artistically and textually is the Hofer Bestiary.[56] This manuscript, whose illustrations were used as models, is unusual in that its text shows a relationship both to the normal DC contents and to H (Book II of Pseudo-Hugo). Almost in the middle of this manuscript are added chapters that can be divided into two groups (ch. 10-12, 13-15) which approximate the order of these chapters in H.[57] The contents of the normal DC text (from Wilhelm's edition) and of the Hofer Bestiary will be listed after an enumeration of the present known illustrated manuscripts.[58]

> Morgan 832, f. 2 - 10v. XII cent. (Formerly Göttweig 101.)
> Vienna, lat. 1010, f. 65-73v. XII cent.
> Munich, lat. 6908, f. 78-85v. XIV cent. (Ends with Salamandra and Ceraste.)
> Épinal 58 (209), f. 70v.-75. XII cent.
> B.N., lat. 10448, f. 118v.-121v. XIII cent. (Small, crude drawings. Moralizations often omitted, and several non-*Physiologus* notices inserted throughout the text, but DC order followed in general.)
> B.M., Sloane 278, f. 44-57. XIII cent. (Preceded by *Aviarium*, f. 1-43v. DC text treats no birds.)
> Hofer Bestiary, f. 60-89. XIII cent. (Preceded by *Aviarium*, f. 17-60.)

---

[55] One example of such additions is found in a thirteenth century manuscript originating in France, Vienna, lat. 303, where the following chapters are included: De mustela (et aspide), de pavone, de corvo, de sale (Menhardt, *op. cit.*, p. 20).

[56] See p. 31, n. 33.

[57] The manuscript to which the Hofer Bestiary stands closest, since both are preceded by the *Aviarium*, is B.M., Sloane 278, but the latter manuscript does not contain all of the additions found in the former.

[58] The following manuscripts are unillustrated, but I have seen no mention of their belonging to the DC version.

> Uppsala, Universitetsbibliotek C 415, f. 1-15v. XIV cent. (Ex monas. Carthus. apud Erford.)
> Arsenal, lat. 394, f. 172-174v. XIII cent. (Irregular order with many chapters missing. Ascribed to *Iohannis Crisostomis*. Preceded by one folio of the *Aviarium*.)

CONTENTS OF DC

WILHELM EDITION	HOFER BESTIARY
1. De leone	1. Leo
2. De panthera	2. De panthera
3. De unicorni	3. De rinocerote
4. De ydro	4. De syrenis et onocentauris
5. De syrenis et onocentauris	5. De ydro et cocodrillo
6. De hyena	6. De hyena
7. De onagro et simia	7. De onagro et simia[59]
8. De elephante	8. De elephante
9. De autula	9. De aspide
10. De serra	10. De lupo
11. De vipera	11. De canibus
12. De lacerta	12. De cheroboles
13. De cervo	13. De adamante
14. De caprea	14. De concha
15. De vulpe	15. De cete
16. De castore	16. De antula
17. De formica	17. De lacerta
18. De ericeo	18. De serra
19. De aquila	19. De vipera
20. De pelicano	20. De cervis
21. De nicticorace	21. De capra
22. De fulica	22. De vulpe
23. De perdice	23. De assida
24. De assida	24. De castore
25. De upupa	25. De formica
26. De caladrio	26. De erinatio
27. De fenice	27. De salamandra
	28. De mustela
	29. De basilisco
	30. De dracone

The list of contents of the double-natured Hofer Bestiary concludes this general survey of the different Latin versions which in their various ways came from Greek originals or grew from basic Latin texts that were either simplified as in the case of Theobaldus' poem or elaborated as were the Second, Third, and Fourth Family versions. Conscious now of the large number and the surprising

---

Although B.N., lat. 2780, f. 93-113v. (end XII cent.) is often listed as belonging to the DC version and it is indeed inscribed *liber Johannis Crisostomi*, the order and text follow H to some extent, but there are also unusual additions.

[59] In reality these are two separate chapters, each with its own illustration, but since Wilhelm numbers Onagro 7 and Simia 7 A, they are kept together in this list.

variety of Latin manuscripts, of which only the oldest or later the most common have been treated here at all, we probably wonder why this work was so extraordinarily popular. In this connection a recent book has named the *Physiologus*, though which version is uncertain, as among the usual contents of an English mediaeval library, and this is doubtless true of continental libraries also. In answer to this question I would hazard the guess that it was the concrete descriptive element, easily retainable because of its very lack of subtlety, which then recalled the Christian dogma or moral exhortation that the reader was summoned to heed.* Thus two areas of man's nature were appealed to by the *Physiologus*: the timeless curiosity about the strange appearance and habits of animals both familiar and unknown (though probably not considered fabulous by many pre-Renaissance readers), and the more specifically mediaeval concept of regarding the external world as a manifestation of the eternal realm of the creator of all things, God. Realizing that a work so attractive in conception and of already proven success could interest an even larger public than those whose reading was largely in Latin, some Anglo-Norman poets —though the word rhymers better describes their talent— in the twelfth and thirteenth centuries translated the *Physiologus* into French.

CHAPTER III

TRADITIONAL FRENCH BESTIARIES

In the one hundred years following the first quarter of the twelfth century the already ancient tradition of the Latin *Physiologus* entered the province of the French vernacular and thus found a far wider audience. During this period were translated the four major bestiaries with which we are concerned in this study: those of Philippe de Thaon, Gervaise, Guillaume le Clerc, and the two versions written in all probability by Pierre de Beauvais. The word "bestiary" is not precisely the correct one to apply to these compositions, since with one exception all were translated not from the large *bestiarius* with its extensive borrowings from Isidore, but rather from the older *Physiologus* of thirty-seven chapters with fixed Isidorean additions. However, the word *bestiaire* is the only term that has long been used to describe the French works. Philippe de Thaon in the oldest French translation writes thus:

> Philippes de Thaün
> En franceise raisun
> At estrait Bestiaire
> Un livre de gramaire,   (1-4)

Apart from the relatively short work by Gervaise which is based on the version known as the *Dicta Chrysostomi*, the three longer bestiaries have as their immediate ancestor a Latin manuscript of the B-Is version. Their close relationship is everywere evident—in a translation which is at times almost literal and in the occasional use of the Latin rather than the French name for an animal or bird. Guillaume le Clerc more than once admits that the French equivalent of a certain animal's name is unknown to him, as in the example

of the Ibis, which he wrongly calls Ibex as do many Latin manuscripts:

> Un oisel est, onc ne fu tex,
> Qui en latin a non ybex;
> Son non en romanz ne sai mie,   (1171-1173)

The appearance of the French manuscripts is generally though
not inevitably one of relative simplicity and even occasional carelessness in comparison with the majority of Latin manuscripts. The
French versions were evidently made for a public less desirous or
demanding in the matter of elegant manuscripts, and as a result
the bestiary in translation was never a de luxe product like its
Latin predecessors or contemporaries.

Certain bestiaries composed in Mediaeval French or in the
dialects of Mediaeval France have been excluded from this study
for one main reason among many — their nature differed sufficiently
either as a whole or in detail to divert these tributaries from the
mainstream of the *Physiologus* tradition. It is felt, however, that
brief mention should at least be made of them, for each contains
something of interest to the student of the bestiary and individual
passages in one often help to explain obscure points in another.

Probably the most popular of all French bestiaries, to judge by
the number and quality of extant manuscripts, was the *Bestiaire
d'Amour* composed in the mid-thirteenth century by Richard de
Fournival, who replaced the habitually ponderous Christian moralization by a light and clever lover's plea for his lady's attention.[1]
Numerous affinities with the long version of Pierre de Beauvais'
bestiary indicate that this work was the principal but not the unique
source of the secular composition, and the latter in turn served as
the basis from which was drawn the bare Cambrai Bestiary, where
religious didactic elements of any sort are almost entirely sup-
pressed.[2] Similar to the Cambrai Bestiary in content and in the

---

[1] A complete critical edition of this work with full bibliographical details
has recently appeared: Cesare Segre (ed.), *Li Bestiaires d'Amours di Maistre
Richart de Fornival e li Response du Bestiaire* (Milan, 1957). All allusions
in this present study to the *Bestiaire d'Amour* will be to this edition.
Previously the only reliable source available for consultation was a non-
critical edition by John Holmberg, *Eine mittelniederfränkische Übertragung
des Bestiaire d'Amour sprachlich untersucht und mit altfranzösischem Pa-
ralleltext herausgegeben*, Uppsala Universitets Årsskrift (Uppsala, 1925).

[2] Edward B. Ham (ed.), "The Cambrai Bestiary", *Modern Philology*
XXXVI (1939), 225-37. The manuscript is Cambrai 370, f. 176v.-178v.

omission of allegorical interpretation is the Provençal version entitled *Aiso son las naturas d'alcus auzels e d'alcunas bestias.*[3] Preserved in only one manuscript of the late thirteenth or early fourteenth century (B.N., fr. 1951) and in one sixteenth century book is an illustrated, rhymed *Bestiaire d'Amour* of 3718 lines.[4] This anonymous poem, although it has some elements found in Pierre de Beauvais and not in Richard de Fournival, seems to be for the most part an imitation of the latter's prose.[5] Finally there exists in a single fifteenth century manuscript at Trinity College, Dublin (C. 5.21, f. 49-70) a moralized bestiary written in the Waldensian dialect and entitled *De las Propriotas de las animanças.*[6] The fifty-four chapters of this bestiary are divided into Birds, Beasts, Fish, and Serpents. The last work to be noted is not properly a bestiary, but knowledge of it is of great value in studying individual descriptions of bestiary animals. The final seventy chapters of Book I of Brunetto Latini's *Li Livres dou Trésor* treat Fish, Serpents, Birds, and Beasts.[7] The closeness in wording in endless examples between the French composition and the Latin of a manuscript of the Second Family leads one to think that Brunetto in many instances did no more than translate the expanded bestiary. But this work is taking us far in time and space from the earliest French translation of the old Latin *Physiologus,* that of Philippe de Thaon.

PHILIPPE DE THAON

The oldest French bestiary and the one closest to the Latin *Physiologus* in some ways is the *bestiaire* of 3194 lines composed

---

[3] Carl Appel, *Provenzalische Chrestomathie* (Leipzig, 1895), pp. 201-4.

[4] Arvid Thordstein (ed.), "Le Bestiaire d'amour rimé: poème inédit du XIIIe siècle", *Etudes Romanes de Lund,* II (1941).

[5] *Ibid.,* p. xxvi. On pp. xix-xxvi Thordstein has listed parallel and dissimilar passages in this work and that of Richard de Fournival.

[6] Alfons Mayer (ed.), "Der waldensische Physiologus", *Romanische Forschungen,* V (1890), 392-418. There has recently been published what is apparently the Latin source (with a few minor differences) of the Waldensian Bestiary although this relationship is not mentioned in the Introduction by J. I. Davis. This is a facsimile edition of a bestiary illustrated with woodcuts and printed between 1508 and 1512 in the Piedmontese town of Mondovi; it is entitled *Libellus de Natura Animalium* (London: Dawson's of Pall Mall, 1958).

[7] Francis J. Carmody (ed.), "Brunetto Latini, *Li Livres dou Trésor*", *California University Publications in Modern Philology,* XXII (1938), 127-71.

by the Anglo-Norman poet, Philippe de Thaon.[8] In the Latin prologue and in the corresponding opening lines in French, Philippe
dedicates this work to Aelis de Louvain, the second wife of Henry I
of England. Since their marriage took place in 1121, it is assumed
that the bestiary was written within a few years of this date.[9] In
the manuscript located at Oxford the recipient of the dedication
has been changed to Alienor, Henry II's wife; this second dedication must date from shortly after 1152.[10] The other known works
by Philippe are a *Comput,* an ecclesiastical calendar written around
1119, and one, if not two, lapidaries.[11]

Three manuscripts of Philippe's bestiary are in existence today:

> L London, B.M., Cotton Nero A. V, f. 41-82v. This
> manuscript dates from the second half of the twelfth
> century, and was once in the possession of the Cister-
> cian m o n a s t e r y of Holmcultram in Cumberland.[12]
> Spaces were left for illustrations which were never
> filled in.

---

[8] Emmanuel Walberg (ed.), *Le Bestiaire de Philippe de Thaün* (Paris
and Lund, 1900). The basis for this edition is B.M., Cotton Nero A.V.

[9] Alexandre H. Krappe, "The Historical Background of Philippe de
Thaün's Bestiaire", *MLN,* LIX (1944), 326. For information about the
family of Philippe de Thaon, see G. L. Hamilton's review of Ch.-V. Langlois'
La Vie en France au moyen âge, Vol. III, *La Connaissance de la nature
et du monde* (Paris, 1927), which appeared in *Speculum,* IV (1929), 112-13.

[10] Langlois' hypothesis (*op. cit.,* p. 4) that Pierre's bestiary might have
existed in two versions seems valid when the strange composition of the
end of the oldest and longest complete manuscript is considered. At line 2889
of the London manuscript the poet, after writing thus far in six-syllable
rhymed couplets, announces at the end of the section on Fire Stones that
he will change his meter; henceforth the verses have eight syllables. After
a gap the poem continues with a notice on the Diamond, the Twelve Stones
of the Apocalypse, and finishes with the Pearl, into which description some
verses on the Beryl have been mingled. The shorter Oxford manuscript,
dedicated to Alienor, after the section on the Owl concludes with sixteen
lines on the Pearl.

[11] Langlois, *op. cit.,* pp. 6-11.

[12] In his description of this manuscript Walberg states that a note by
the copyist indicates that the manuscript was written in the monastery of
Holmcultram in Cumberland. It has recently been pointed out, however,
that the writing of the ex-libris on f. 82v., *Liber sancte marie de holm-
coltrum,* is not contemporary with the scribe's handwriting and might be
later by a century. See Fr. John Morson, O.C.R., "The English Cistercians
and the Bestiary", *Bulletin of the John Rylands Library Manchester,* Vol. 39,
No. 1 (September 1956), 168-69.

O   Oxford, Merton College 249, f. 1-10. The manuscript seems to have been written in England in the thirteenth century.[13] After the section on the Owl the text includes sixteen lines on the Pearl (ll. 3063-3079) and ends apparently incomplete. Forty-eight drawings exist, for the most part poorly executed.[14]

C   Copenhagen, Royal Library 3466, f. 3-51. This unfinished manuscript stops at l. 1928 in the section on the Aspidochelone. It contains twenty-eight interesting drawings. Possibly this thirteenth or fourteenth century manuscript was copied at the former Priory of Saint Martin des Champs in Paris.

An unusual feature of the text is that all three manuscripts contain not only a Latin prologue but also Latin rubrics which either state the contents in a terse manner or give directions to the illustrator,[15] e. g., "Hic leo pingitur et quomodo capit animalia per circulum". The French formula for this is:

> Ço mustre la peinture,
> Si est dit par figure.   (107-108)

The authorship of these rubrics has puzzled scholars who have examined them. Mann, one of the first to study the works of Philippe de Thaon,[16] and Langlois with him,[17] thought they were not the work of the poet; but the editor of the *bestiaire,* Walberg, considered that the question was not yet settled, and in particular could conceive

---

[13] Walberg, *op. cit.,* p. vi.

[14] The manuscript is written in double columns in the form of prose. There seems to have been no cooperation between the copyist and the illustrator since the former often did not leave sufficient space for the drawings, which were consequently squeezed into the margins.

[15] In both the Oxford and the Copenhagen manuscripts it appears at times that neither the copyist nor the illustrator understood Latin. In O the Latin directions occasionally are buried in the French text with no room at all left for the required illustration, and in C, after a drawing of the Stag appears following the Hydrus allegory, the illustrations usually occur one space before they correctly should. They are also poorly placed in the large spaces allotted to them.

[16] Max Friedrich Mann, "Der Physiologus des Philipp von Thaün und seine Quellen", *Anglia,* VII (1884), 435-37. Mann's study includes pp. 420-68 in Vol. VII, and pp. 391-434, 447-50 in Vol. IX.

[17] Langlois, *op. cit.,* p. 15, n. 3.

of no one other than the author as writing the Latin prologue.[18]
Because of the close affinity between the brief Latin summaries
and the French text, I am inclined to think that Philippe composed
them both, although I can find no explanation for the author's
concerning himself with indications to the artist.

Philippe de Thaon's bestiary contains thirty-eight chapters,[19]
which are divided as the Latin prologue states:

> Liber iste Bestiarius dicitur
> Quia in primis de bestiis loquitur
> Et secundario de avibus,
> Ad ultimum autem de lapidibus.

Chapters 1-23 treat of Beasts, beginning as always with the Lion.
The Ostrich is also included here — perhaps because the Latin
source reads "Item est animal quod dicitur asida". Chapters 24-34
describe Birds, but the Eagle, which usually introduces the section
on Birds, is preceded by the Partridge. Lauchert has rightly suggested
that this is possibly a remnant of the Latin order (as is seen in B)
where the Partridge follows the Aspidochelone.[20] The final four
chapters present Stones, including the Twelve Stones of the Apocalypse, not found elsewhere in French but noted in at least two
Latin manuscripts of the Second Family. Since the order of chapters
within the three divisions of this work is unlike any other Latin
or French bestiary — it bears no resemblance to the *Dicta Chrysostomi* — this might partially account for some of the startling
statements made about Philippe's composition in relation to its
obvious Latin prototype.[21] Although a Latin manuscript with the
same arrangement as the French version might have existed, it seems
equally probable that Philippe created his own order. That the
notices were grouped so as to illustrate certain theological types
(Christ, the devil, man) is apparently not the explanation for the

---

[18] Walberg, *op. cit.*, p. ci.

[19] Walberg combines Adamas (36) and *Les douze pierres* (37) to make
a total of thirty-seven.

[20] Lauchert, *op. cit.*, p. 129.

[21] One of the last words, though already twenty years old, to be written
on this relationship is that of Carmody, "De Bestiis et Aliis Rebus and the
Latin Physiologus", *Speculum*, XIII (1938), 154, who writes in a critical
list of extant versions of the *Physiologus*, "Q - Philippe de Thaon, *Bestiaire*,
in T. Wright, *Popular Treatises on Science* (London, 1841), pp. 74-131; not
related to the Latin *Physiologus*". Carmody sees this work as deriving from
the Greek, like Bern 318 (C).

unusual order, since Philippe's allegories, although occasionally showing a distinctly individual interpretation, on the whole do not greatly differ from the traditional moralizations.[22]

As sources Philippe de Thaon cites *Phisiologus* fourteen times, *bestiaire* or *bestiaire, un livre de gramaire* nine times, *Ysidre* eight, and *escripture* often. Sometimes both *bestiaire* and *Phisiologus* are mentioned in the same passage, as in the Phoenix where not only does the name *Ysidorus* appear but

> De lui dit Bestiaire
> Chose ki mult est maire,
> E Phisiologus
> De lui dit uncor plus.   (2247-2250)

Lauchert has explained this by the fact that the Latin book was usually called *Bestiarius*, while the contents often begin with the formula *Physiologus dicet...*[23] These references to sources lead to a question which troubled early students of Philippe de Thaon, although part of their difficulty can be attributed to lack of published texts at the time they were writing: did the Latin work from which Philippe drew his material already contain the citations from Book XII of Isidore's *Etymologiae*, or did he consult this work directly? Cahier believed in the existence of an already interpolated Latin bestiary, although he had not seen one,[24] whereas Lauchert declared that such a hypothesis could be proved only by the discovery of a manuscript containing Isidore's additions.[25]

As shown in Chapter II of this study, not a few manuscripts of this type—called here B-Is—have been found. Bodl., Laud. Misc.

---

[22] Although Walberg states that Pierre is "archaic" in his choice of types and allegories (*op. cit.*, pp. 18, 19), he was probably imitating his model.

[23] Lauchert, *op. cit.*, pp. 132-33. However, in two of the three cases where both *Phisiologus* and *bestiaire* are cited in the same article, the former word corresponds exactly to its position in the Latin text, and the mention of the *bestiaire* in the allegory is no more than a metrical filler (e.g., l. 1773 and l. 2128). But the pattern in the above quoted passage on the Phoenix is more revealing. *Ysidorus* is named in the middle of the description drawn precisely from his work: the Isidorean account ends with the reference to *bestiaire* (l. 2247); and the true *Physiologus* text begins by quoting from *Phisiologus* (l. 2249).

[24] Cahier, *op. cit.*, II, 97.

[25] Lauchert, *op. cit.*, p. 133, n. 1. Mann erroneously linked this bestiary with both A (Brussels 10074) and B (Bern 233), *Anglia*, IX, 433.

247, which James chose to illustrate the First Family, is one of these, while the almost identical B.M., Royal 2 C. xii, which Mann linked with the bestiary of Guillaume le Clerc, is another. Walberg unobtrusively reached the conclusion of a relationship between Philippe de Thaon's *bestiaire* and the latter Latin manuscript.[26] Indeed, the contents correspond so closely on the whole, although the internal order of the various chapters might at times differ, that we can surely say a manuscript of the B-Is version served as Philippe's model with the few exceptions that will be noted further on. One example of the countless similarities that occur is found in the chapter on the Ant. After recording the three natures of the Formica, the Latin of Laud Misc. 247 says, "Ethimologia. Formica dicta quod ferat micas farris." This Philippe renders, and not correctly this time:[27]

> Uncor de furmi dit
> Ysidre en sun escrit
> E bien mustre raisun
> Pur quei furmi at nun:
> Forz est e porte mie,...   (1031-1035)

After a few sentences the Latin says, "dum pluit super frumentum eius, totum eicit". Philippe writes,

> S'il plot sur sun furment
> Gete le fors al vent.   (1047-1048)

When the Latin of the manuscript reads, "Dicuntur et in Ethiopia esse formicas ad formam canis", Philippe says,

> Uncor Ysidorus
> D'altre furmi dit plus:
> En Ethiopie en sunt
> Ki del grant del chien sunt.   (1053-1056)

His account of the Ethiopian Ant follows closely the Latin text, which in this case is much more developed than Isidore's. The

---

[26] Walberg, *op. cit.*, p. xxx, n. 2.

[27] The only other example discovered of Philippe's deviating in etymology from Isidore is in the account of the Partridge:

> Perdix d'oisel est num,
> E pur ço at tel num
> Que pert sa nureture,   (1959-1961)

existence of the extended version of the Ethiopian Ant in both the Latin and the French precludes a direct borrowing from Isidore on the part of Philippe.

Another item points to a relationship — in this case artistic — between one manuscript of Philippe's bestiary and a manuscript similar to Laud Misc. 247. In the Latin section on the Hydrus, after the *Ethimologia* is cited and passages quoted from Isidore, the Crocodrillus (sic) is described as being "solus autem pre animalibus superiora oris movet". This peculiarity is not mentioned by Philippe, but in the illustration of the Copenhagen manuscript (Pl. V, Fig. 2 a), the dog-like Crocodile's head is on upside-down in a drawing similar to that of Laud Misc. 247.

There are also obvious differences between the Latin B-Is versions and the French which speak for Philippe's forgetfulness or his fancy, or for a different source for certain details. Where, for instance, did the author find the story of the Dog's pulling up the Mandrake? The normal B-Is manuscript does not include the important, early account of the Elephant's sleeping against a tree, which Philippe records. In the Latin the Siren is not a heterogeneous bird-fish as in the French, but only a bird; nor does Philippe's puzzling first chapter on the Lion with its numerous illustrated traits find an exact parallel in any Latin *Physiologus* now known. It is thus evident that many points remain to be solved in this oldest French bestiary.

In order to indicate the striking similarity of names and general content, the chapters of Philippe de Thaon's bestiary are here listed with the corresponding notices from Bodl., Laud Misc. 247:

1.	Leun	1.	Leo
2.	Monosceros	16.	Monoceros
3.	Pantere (et Dragun)	24.	Panthera (et Draco Maior)
4.	Dorcon	20.	Dorcon
5.	Ydrus (et Cocodrille)	19.	Hidrus (et Crocodrillus)
6.	Cerf	31.	Cervus
7.	Aptalon	2.	Antalops
8.	Furmi	11.	Formica
9.	Onoscentaurus	12.	Onocentaurus (et Syrena)
10.	Castor	17.	Castor
11.	Hyena	18.	Hiena
12.	Mustele	27.	Mustela
13.	Assida	29.	Assida
14.	Sylio (Salamandre)	32.	Salamandra
15.	Serena	12.	Syrena (et Onocentaurus)
16.	Elefant	35.	Elephantus
17.	Aspis	28.	Aspis

18. Serra	4. Serra
19. Heriçun	13. Herinacius
20. Gupil	15. Vulpis
21. Onager	21. Onager
22. Singe	22. Simia
23. Cetus	25. Aspidochelone (Cetus)
24. Perdix	26. Perdix
25. Aigle	8. Aquila
26. Caladrius	5. Caladrius
27. Fenix	9. Fenix
28. Pellicanus	6. Pelicanus
29. Colum (et Peredixion)	33. Columbe. (34). Peredixion
30. Turtre	30. Turtur
31. Huppe	10. Hupupa
32. Ibex	14. Ibex
33. Fullica	23. Fulica
34. Nicticorax	7. Nicticorax
35. Turrobolen	3. Terobolem
36. Adamas	37. Adamans
37. Douze pierres[28]	
38. Union	38. Mermecolion

To conclude this brief survey of the first French bestiary, a word in defense of the style and subject matter of Philippe de Thaon's composition seems appropriate. From Mann, who called the *bestiaire* "eine müssige anhäufung von ungeniessbaren naturwissenschaftlichen armseligkeiten, in denen sich das mittelalter so sehr gefiel",[29] through Paul Meyer, who states about Philippe, "Ses écrits ont une certaine importance pour l'histoire de la langue et de la versification, mais c'est leur seul mérite, car ils ne se distinguent ni par l'originalité des idées, ni par la distinction du style",[30] to a verbose diatribe by Langlois,[31] all contain a measure of truth. But it should be remembered, as Paul Meyer did concede, that no more than a translation was proposed, and that this is an early work in a language still groping to express itself. All of these scholars appear to miss the excitement inherent in the fact that a tradition already ancient and rich had now entered the vernacular to become widely read and known in the next century and a half.

---

[28] For a discussion on the inclusion of these stones, see Langlois, *op. cit.*, pp. 3-11. Philippe omits the traditional Latin chapter entitled Amos propheta which precedes Adamas.

[29] Mann, *Anglia*, VII, 447.

[30] Paul Meyer, "Les Bestiaires", *Histoire littéraire de la France*, XXXIV (1914), 368.

[31] Langlois, *op. cit.*, pp. 12-13

GERVAISE

The short (1280 lines) rhymed *Bestiaire* of Gervaise is thought to have been written at the beginning of the thirteenth century.[32] In the prologue the author states:

> Gervases ...
> Vuet .l. livre en roman traite.   [r]
> Li livres a non Bestiaire.
> A Barbarie est [en] l'armaire
> Li latins qui mult est plaisanz;
> De illuec fu estraiz li romanz.
> Celui qui les bestes descrist
> Et qui lor natures escrit
> Fu Johanz Boche d'or nommez,
> Crisothomus rest apelez.   (32-40)

The Barbarie mentioned in these lines is thought to be Barberie or Barbery, a Cistercian abbey of the diocese of Bayeux founded in 1176, and although three men by the name of Gervasius have been noted in this region, there is nothing that with any certainty connects one of them with this work.[33] The bestiary exists in a single manuscript, B.M., Add. 28260, f. 84-100v., from the second half of the thirteenth century. There are small, crude drawings in the margins up to that of the Raven, which is within the body of the text (f. 93), but afterwards the spaces in the text are empty.

The attribution of this work to John Chrysostom would inevitably lead to a comparison of its contents with the version known as the *Dicta Chrysostomi*.[34] Lauchert, who unwittingly omitted any reference to Gervaise in his standard work, belatedly made this comparison and found a decided similarity as well as some differences between Gervaise's composition and the various manuscripts of the *Dicta Chysostomi*.[35] At present none of the known Latin manuscripts of this version contains exactly the same order of contents as the French translation. The principal details in which the two works diverge are that the Siren is a woman-fish in Gervaise while it retains the earlier form of a woman-bird in the Latin; the father Pelican pierces itself in the French version, whereas it is the

---

[32] Paul Meyer (ed.), "Le Bestiaire de Gervaise", *Romania*, I (1872), 420-43.

[33] *Ibid.*, pp. 421-22.

[34] See the section entitled *Dicta Chrysostomi* in Chapter II of this study.

[35] F. Lauchert, "Nachträgliches zum Physiologus", *Englische Studien*, XIV (1890), 128.

mother in the Latin; the fire that burns the Phoenix's nest comes
from stones in Gervaise, and from the sun in the Latin; and finally
the odd description of the Saw Fish (*Sarce*) in the French. Both
the bestiary of Gervaise and the *Dicta Chrysostomi* have the same
unusual way of presenting the three characteristics of the Snake in
the same chapter with the Viper.

Lest it be thought that Gervaise changed or enlarged the manuscript which he was translating, let him vindicate himself:

> Ici fenist li Bestiaires.
> Plus n'en avoit en l'essenplare
> Et de mentir seroit folie.
> Qui plus en set plus vos en die!
> Gervaises qui le romain fit
> Plus ne trova ne plus n'en dit.
> CI FENIST LI LIVRE DES BESTES;
> DEX NOS GART NOS BIENS ET NOS TESTES!   (1273-1280)

The contents of the *Bestiaire* of Gervaise are listed below with
the typical chapters of the *Dicta Chysostomi* for comparison.

1. Lion	1. Leo
2. Panthere	2. Panthera
3. Unicorne	3. Unicornis
4. Idres et Cocadrile	4. Ydrus
5. Sereine. 6. Centaurus	5. Syrene et Onocentaurus
7. Hyene	6. Hyena
8. Singe	7. Onager et Simia
9. Elephant	8. Elephas
10. Antule	9. Autula
11. Serpent (et Vuivre)	10. Serra
12. Corbeau	11. Vipera
13. Vurpil	12. Lacerta
14. Castor	13. Cervus
15. Eriçon	14. Capra
16. Formi	15. Vulpis
17. Aille	16. Castor
18. Caradrius	17. Formica
19. Pellicanus	18. Ericeus
20. Perdriz	19. Aquila
21. Chamoi	20. Pellicanus
22. Hupe	21. Nocticorax
23. Phenix	22. Fulica
24. Cerf	23. Perdix
25. Tortre	24. Assida
26. Sarce	25. Upupa
27. Belete	26. Caladrius
28. Aspis	27. Phoenix
29. Ibis	

GUILLAUME LE CLERC

The most artistically composed and the longest (3426)[36] of the rhymed French bestiaries was written by another Norman, Guillaume, surnamed le Clerc or le Normand.[37] The popularity of this version is attested by the fact that there are extant some twenty-three manuscripts copied in England and France, dating from the thirteenth to the fifteenth centuries, and of these the large majority are carefully illustrated.[38] The present state of the manuscripts is the following (Reinsch's lettering is used):

A - B.M., Egerton 613, f. 31-59. Illus. Mid-XIII cent.

B - B.N., fr. 14969, f. 1-72v. Illus. XIII cent.

C - B.N., fr. 2168, f. 188v.-209v. Unillus. Second-half XIII cent.

D - B.N., fr. 25406, f. 1-30. Spaces unfilled. End XIII or XIV cent.

E - B.N., fr. 14964, f. 118-181v. Illus. End XIII cent.

F - B.N., fr. 1444, f. 240-257. Illus. End XIII cent.

G - B.N., fr. 14970, f. 1-34v. Illus. XIII cent.

H - B.N., fr. 24428, f. 53-78v. Illus. XIII cent.

I - B.N., fr. 25408, f. 70v.-106v. Unillus. XIII cent.

K - B.N., fr. 902, f. 137-159. Unillus. XIV cent.

L - B.N., fr. 20046, f. 1-36. Partially illus. XIV.

---

[36] The Reinsch edition of Guillaume's bestiary (see below) contains a total of 4174 lines. This length is explained by the fact that Reinsch used as the basis for his edition B.M., Egerton 613, which, like several of the manuscripts, goes on after the final bestiary chapter on the Diamond to include additional compositions by Guillaume. These are the parables of the talents and of the workers in the vineyard, and some brief reflections on the three enemies of man (the devil, the world, and the flesh).

[37] Robert Reinsch (ed.), *Le Bestiaire, Das Thierbuch des normannischen Dichters Guillaume le Clerc* (Leipzig, 1890). An English translation of this bestiary exists by George C. Druce, entitled *The Bestiary of Guillaume le Clerc*. This book was printed for private circulation by Headley Brothers, Invicta Press, 1936.

[38] Paul Meyer's additions in his article "Les Bestiaires", *op. cit.*, p. 375, n. 5, to the manuscripts listed by Reinsch, *op. cit.*, pp. 13-31, bring the list to twenty-one; others have been added by George C. Druce, "The Elephant in Medieval Legend and Art", *Archaeological Journal*, LXXVI (1919), 18.

M - B.M., Royal 16 E. viii, f. 2-71v. Illus. XIII cent. Miss-
     ing since 1879.

N - B.M., Cotton Vesp. A. vii, f. 4-33. Illus. XIV cent.

O - Barrois 11, formerly belonging to Lord Ashburnham.
     Now B.N., Rothchild IV.2.24, f. 140-163. Unillus.
     XIV cent.

P - Bodl., Douce 132, f. 63-81v. Illus. XIV cent.

Q - Vatican, Regina 1682, f. 4-26v. Unillus. XIV cent.

R - This designation appears to belong to B.N., fr. 25406.

S - Berlin, Hamilton 273. Now Camb., Fitzwilliam Mu-
     seum, J.20, f. 45-73. Illus. XIV cent.

T - Lyon, Palais des Arts 78, f. 36-58. Illus. XIII cent.

U - Phillipps 4156. XIII cent.

  - Camb., Trinity Coll., 0.2.14(2), f. 32v.-61v. Illus. XIII
     cent.

  - Camb., Fitzwilliam, McLean 123, f. 20-66. Illus. *ca.*
     1300.

  - Bodl. 912, f. 1-12. Illus. Early XIV cent.

  - Paris, Arsenal 2691, f. 62-95v. Spaces unfilled. XV
     cent.

The illustrations in Bodl. 912 seem in many instances to
be poor copies of B.M., Egerton 613, as do some of Fitzwilliam, McLean 123.

The *Bestiaire,* or *Bestiaire divin* as the title appears on some
manuscripts, was addressed to a certain Raoul, who remains unknown, around 1210 or 1211. The basis for this date is found in
the prologue where the author says:

> Voelt Guillame en romanz escrire
> De bon latin, ou il le troeve.
> Ceste ovraigne fu fete noeve
> El tens que Phelipe tint France,
> El tens de la grant mesestance,
> Qu'Engleterre fu entredite,   (8-13)

Later (ll. 2707 ff.) he adds that the work was written two years
after England was put under interdict (this had been ordered by
Pope Innocent III on March 23, 1208).

That the reader should profit from the moral lesson expounded in his bestiary is Guillaume's stated aim;[39] I wonder, however, if it was not rather the accounts of the animals and the occasional rather elegant illustrations on the manuscripts which proved more attractive to the number of people who wished copies than the "essample prendre / De ben faire e de ben aprendre". Guillaume's allegories usually repeat with some embellishments the traditional material found in the Latin of the B version, but on occasion he will include a personal plea on the necessity of faith and good works, and at one point, in the allegory of the Asp, there appears an extended passage on the evils of wealth in the form of an exemplum telling of a rich man who casts his gold into the sea. Guillaume's preocupation with the moral import of his work found an echo in two of the illustrators of his bestiary who, in the only examples now known, consistently illustrated the allegory as well as the animal or bird in question.[40] However, without a knowledge of the text, the allegorical scenes depicted would be quite difficult to understand. The individual flavor which Guillaume imparted to his bestiary, in contrast with Philippe de Thaon's impersonal composition, is evident also in his introducing occasional literary and contemporary allusions — rare as they are — as, for exemple, to Arthur, Charlemagne, and Ogier (l. 565), to Renart's stealing the hens of Costans de Noés (l. 1308), and the above-mentioned remarks on England's being under interdict while the poet was rhyming his work.[41] In all, Guillaume's octosyllabic poem reads smoothly and has a certain picturesque quality about it.

It seems incontestable today that the Latin source used by Guillaume was none other than a manuscript of the widely spread version which is called in this study B-Is.[42] It was the German scholar Max Friedrich Mann who first showed the dependency of Guillaume's *Bestiaire* on a Latin version whose nature and wording have been preserved in the unillustrated thirteenth century manuscript which has already been referred to many times, B.M., Royal 2 C. xii.[43] In 1914 Paul Meyer readily accepted the relationship

---

[39] ll. 25-36.

[40] These two manuscripts are B.N., fr. 24428 and 14969.

[41] In the parable of the workers in the vineyard Guillaume also mentions Maurice de Sully, bishop of Paris, who died in 1196.

[42] See Chapter II.

[43] Max Friedrich Mann, "Der Bestiaire Divin des Guillaume le Clerc", *Französische Studien*, VI, Heft 2 (Heilbronn, 1888), pp. 35 ff.

demonstrated by Mann,[44] but others previous to him were unconvinced, although their reasons scarcely seem valid today. Reinsch, the editor of Guillaume's bestiary, asserted that Book II of Pseudo-Hugo of Saint Victor had as much in common with the bestiary of Guillaume as the manuscript studied by Mann,[45] but an impartial examination of the manuscripts disproves this statement, for not only is the order — with two exceptions — identical, but many significant details are similar in the French version and in the Latin of Royal 2 C. xii. Another discordant voice was that raised by Max Goldstaub and Richard Wendriner, who stated their belief that Guillaume le Clerc had not used a uniform text as a source throughout.[46] Now that the extensiveness of the B-Is version in recognized, there seems no reason to think that Guillaume's is a composite work. Except for the omission in the French bestiary of the articles on the prophet Amos and on the Pearl, Guillaume's chapters follow the normal order of the Latin B-Is and include in almost all instances the Isidorean additions which had become an integral part of the old B version.

Differing from Philippe de Thaon, Guillaume does not name Isidore as the one from whom he drew much information, although he occasionally refers to *li livres* and to *le bestiaire* as sources. But just as in Philippe's bestiary, which quite naturally it often resembles in content,[47] there are variations in omissions and inventions in the text that might be due to Guillaume or to his model; it is difficult at this point of textual studies to say definitely to which. Some of the differences between Guillaume le Clerc's *Bestiaire* and Royal 2 C. xii are: in the account of the Pelican the father bird pierces its side (the Latin says *mater*); only the B version of the Phoenix story, where India is named as the dwelling place of the bird, is given, and the Phoenix's funeral pyre is lighted by sparks from its beak striking against stone; the Siren is like a fish or a bird, while the Onocentaur is omitted entirely; the Hedgehog collects both grapes and apples on its quills; there is no conclusion to the

---

[44] Paul Meyer, "Les Bestiaires", *op. cit.*, p. 376.

[45] Reinsch, *op. cit.*, p. 68, n. 2.

[46] Max Goldstaub and Richard Wendriner, *Ein Tosco-Venezianischer Bestiarius* (Halle, 1892), p. 93.

[47] Mann rightly attributes their resemblance to the similarity of their source, *op. cit.*, p. 90.

tale of the Ape, which carries the preferred offspring in its arms (this unfinished state also exists in Philippe de Thaon's bestiary); and the Turtle-Dove will sit on nothing green. Minor as these points may seem, there would be satisfaction in being able to account for them.

Below are listed the contents of the *Bestiaire* of Guillaume le Clerc (based on B.M., Egerton 613) and those of the Latin manuscript, B.M., Royal 2 C. xii.

1. Lion	1. De tribus naturis leonis
2. Aptalops	2. De autalops
3. Dous peres	3. De lapide igniferi (terebolem)
4. Serre	4. De serra in mari
5. Caladrius	5. De chelindro, caladrius
6. Pellican	6. De pelicano
7. Nicticorace, freseie	7. De nicticorace
8. Aigle	8. De aquila
9. Fenis	9. De fenice
10. Hupe	10. De huppupa
11. Formi	11. De tribus naturis formice
12. Sereine	12. De sirena et onocentauro
13. Heriçon	13. De herinaceo
14. Ybex	14. De ibice
15. Gupil	15. De vulpe
16. Unicorne	16. De monocero
17. Bevre	17. De castore
18. Hyaine	18. De hiena
19. Idrus, Cocadrille	19. De hidris
20. Buc, chevre	20. De dorcon
21. Asne salvage	21. De honagro
22. Singe	22. De simia
23. (Fulica) - unnamed	23. De fulica
24. Pantere, Dragon	24. De panthera
25. Cetus	25. De duabus naturis aspidis celonis
26. Perdriz	26. De perdice
27. Belette, Aspis	27. De mustela, de aspide
28. Ostrice	28. De assida et strucione
29. Turtre	29. De turture
30. Cerf	30. De cervo
31. Salamandre	31. De salamandra
32. Colom, Paradixion	32. De columbarum naturis
33. Olifant	33. De arbore peredixion
34. Mandragoire	34. De elephanto
35. Aimant	35. De amos propheta
	36. De adamante
	37. De mirmicolion

PIERRE DE BEAUVAIS

The most interesting and puzzling French bestiary is that composed before 1218 by a certain Pierre. Since the oldest manuscript of his prose bestiary is in the Picard dialect, Cahier called him Pierre le Picard, while Gaston Paris limited the name to Pierre de Beauvais.[48] Almost every aspect of this work by Pierre presents tantalizing problems, the majority of which have by no means been solved here although some new information is presented.

First among the unusual features of this bestiary is that it exists in two forms: a short version of about thirty-eight chapters whose order and content, as Paul Meyer has indicated,[49] resemble that of the oft-mentioned B.M., Royal 2 C. xii, but with a few additions which escaped the French scholar's attention; and a long version of some seventy-one chapters containing all the material of the short version supplemented by an almost equal number of descriptions drawn from other sources.

The following manuscripts are known to exist.[50]

Long version:

    P - Paris, Bibl. de l'Arsenal, fr. 3516, f. 198v.-212v. XIII cent. Illus.

    Mon - Montpellier, Bibl. de la Faculté de Médecine, H. 437, f. 195-250 (end missing). XIV cent. Illus.

    V - Vatican, Reg. 1323, f. 2-36. Dated 1475.[51]

    Ph - Phillipps 6739, f. 1-50. Late XIII cent. Illus.[52]

---

[48] Cahier, *Mélanges d'archéologie*, Vol. II, 106-232; Vol. III, 203-88; Vol. IV, 55-87. The precise form of the name is given by Gaston Paris in *Romania*, XXI, p. 263.

[49] Meyer, "Les Bestiaires", *op. cit.*, pp. 381-90.

[50] The letters P, R, S were assigned by Cahier; the others by the present writer.

[51] This manuscript has previously received only a brief notice by Langlois in *Notices et Extraits des Manuscrits de la Bibliothèque Nationale*, XXXIII, 2 (1889), p. 111. It is written in a fifteenth century hand that is at times difficult to read. At the end of the text on f. 36 the following *explicit* is found: "Explicit le grant bestiaire commencé par moy Jehann Pamir a Pons Sainte Maxence et parchevé au chasteau de Bouillencourt le lundi XVIII° jour de septembre mil iiii c lxxv." The first place referred to is in all probability Pont Sainte-Maxence (Oise) in the north of Ile-de-France, while the castle could be located either at Bouillancourt (Somme, arr. de Montdidier) or at Bouillancourt-en-Sery (Somme, arr. d'Abbeville).

[52] This number was incorrectly printed in Meyer's article on French bestiaries as 6730. The description in the *Catalogus librorum manuscriptorum*

Short version:

R - Paris, B.N., fr. 834, f. 39-48v. XIV cent.

S - Paris, B.N., fr. 944, f. 14-34v. XV cent.

L - Paris, B.N., nouv. acq. 13251 (former La Clayette MS.), f. 22-31. XIII cent. Illus.

Ma - Malines, Bibl. du Séminaire 32, f. 1-23. XV cent.[53]

In order to present clearly the contents of each version, the titles in a representative manuscript of each type will be listed.

Short version	*Long version*
R - B.N. 834[54]	V - Vatican, Reg. 1323
1. Du lyon	1. Du lyon premier
2. Del antula	2. De l'autula
3. De ii pierrez ardanz	3. De la sarre
4. De la serre	4. Du turobolem
5. Del caladre	5. Du caladres
6. Du pelican	6. De la guivre
7. Du niticorax	7. Du pelican
8. De l'aigle	8. Du tigre
9. Del fenis	9. De la grue
10. (De la hupe)[55]	10. Du voultre
11. Du formi	11. De l'aronde

---

in bibliotheca d. Thomae Phillipps, bart. (1837) reads: 6739 Le Livre apelé Bestiaire, translatei de latin en Roumans par Pierre ki le fist par le commandement de l'Evesque Philipon. I examined this manuscript in London and found it to be a complete manuscript of the long version ending with the bird called Rafanay.

[53] The existence of this manuscript was brought to my attention through the kindness of M. Omer Jodogne of Louvain, who is preparing an edition of Pierre de Beauvais' *Bestiaire*. Its contents appear to be the same as R and L, to judge by the short description in the *Catalogue général des manuscrits des bibliothèques de Belgique.* Vol. IV, *Catalogue des manuscrits du grand séminaire de Malines* (1937). It ends on f. 23 thus: "...ataindre a ce quil convoitent et ce pierdent quil guerpissent. Chi fine li bestiaire."

[54] The names of the animals in the rubrics are preceded in many instances by the phrase *Selonc les proprietez du (lyon)* or *Ci parle des proprietez de...* It should be noted that the text of the short version differs at times from the long text and variants published by Cahier. For example, in the chapter on the Aspis in S Cahier fails to mention that the author, following the Latin of the B-Is version, names various kinds of Asps — *stala, prialis, emorois,* among others.

[55] There is no rubric for the Hoopoe, which is a continuation of the text on the Phoenix.

12. De la seraine
13. Du heriçon
14. Del ÿbex
15. Du gourpil
16. Del unicorne
17. Del castre
18. Del hÿene
19. Del ydre
20. De la chievre
21. De l'asne sauvage
22. Du singe
23. Del felica (text: fulica)
24. De la penthere (text: panthere)
25. De la coine
26. De la pertris
27. De la mostoile
28. De l'asida
29. De la tortre
30. Du cerf
31. De la salemandre
32. De la tanrine coulor[56]
33. Du coulon
34. Del olifant
35. De la chievre sauvage (Amos)
36. Del aimant
37. Du leu
38. Du chien

12. Du voltours
13. De l'asepic (text: aspis)
14. Du crinon ou gresillon
15. Du corbiau
16. De l'arpie
17. Du rosignol
18. Du paon
19. De l'espec
20. De l'alerion
21. De l'aisgle
22. Du niticoraux[57]
23. De la seraine
24. De la huppe
25. De la vasche[58]
26. Du fenis
27. Du papegay
28. Des fourmis
29. De l'ostruche
30. Du herison
31. Du ybeux
32. Du goupil
33. De l'aringne
34. Des annes de la mer[59]
35. Du baselicoc
36. Du teris
37. De unicornes
38. Du grifon
39. Du castoire
40. De la hienne
41. De ulica
42. Du cocadrille et de hydre
43. Du dorcon
44. Du centicore
45. De l'asne sauvage
46. Du singe
47. Du signe
48. Du huhan[60]

---

[56] L calls this *Laurine color*. Its origin is the chapter on Columbae (B-Is 32): "Physiologus dicit multis ac diversis coloribus esse columbas, id est color... aurosus..." R. begins: "Tanrine color sénéfie les III Enfans..." In Pierre this notice treats the various colors of the Dove while the following one is on the Peridexion Tree and the Dragon.

[57] In P, although the text is on the Owl, the title is *Cauve sorvis* [sic]. This same confusion occurs also in Queen Mary's Psalter (B.M., Royal 2 B. vii) where on f. 91v. and 93 there is an illustration of the Bat while the correct order would call for the Owl.

[58] P's title is *Argus le vachier*.

[59] P's title: *l'arbre dont li oisel naisent fors et chient jus quant il sont meur.*

[60] P - *huerans*.

49. Du pantere
50. De la pertris
51. De la covie
52. Du asida
53. De la torte
54. De la mesenge
55. Du serf
56. De la salamandre
57. De la taupe
58. Du coulon
59. De l'arbre des coulons[61]
60. De l'olifant
61. Des chievres (Amon li prophe-
    tes)
62. De l'aimant
63. Du lou
64. D'ung poisson qui est appelés
    esynus
65. Des chiens
66. De l'omme sauvage[62]
67. De la merle
68. De l'escoufle
69. Du muscaliet
70. Des quatre elemens[63]
71. De l'orphanay

The variations contained within the prologue are important because of the alleged source cited in some of the manuscripts and
because of the name of the person at whose request this bestiary
was written, since it is on this information that the date of composition rests. The following is Paul Meyer's transcription of P:

Chi comenche li livres des natures des bestes.

Chi commence li livres c'on apele Bestiaire. Et por ce est il
apelés ensi, qu'il parole des natures des bestes, car totes les
creatures que Dex cria en terre, cria il por home, et por prendre
essanple et de foi en eles et de creance. En cest livre translater
de latin en romans mist grant travail et grant cure Pieres, qui

---

[61] P's title: *Del dragon. del arbre de Judée. et del colon.*

[62] P's title: *del Sagetaire et del salvage home*; Mon's title: *Des gens
as cornes ou front.*

[63] P has two chapters here instead of one: *De coi li home est fais et
de sa nature* and *del voltoir. et de un ver con apele lieus.*

volentiers le fist par le commandement l'evesque Philipon Cuers[64] en cui service ne perist mie, car il est espeisse debonaires, laituaires de franchise et confors de guerredon. Et por ce que rime se feit afaitier de mos concueillis hors de verité, volt li evesques que cist livres fust fait sans rime tot selonc le latin que Phisiologes, uns des bons clers d'Athenes traita.[65] Et en tous sens les natures des bestes et des oiseaus a l'entendement des spiriteus cose.

The Philippe to whom reference is made in five of the manuscripts has been identified as Philippe de Dreux, the turbulent bishop of Beauvais who died in 1217.[66] The dedication to Count Robert that exists in the Malines manuscript is doubtless addressed to the same person for whom Pierre wrote his *Mappemonde,* Robert II, Count of Dreux and brother of Philippe.[67] From this grouping of names and dates it appears certain that Pierre's *bestiaire* in both its long and short form existed before 1218, the year in which Robert II died.

----

[64] The dedications in the various manuscripts are as follows: Philipon Cuers - P, V, Mon; Philippe - S, Ph; Conte Robiert - Ma; No name - R, L. According to P. Meyer, "Les Bestiaires", p. 384, n. 3, the name "Cuers" exists nowhere else pertaining to this bishop. Cahier, II, p. 106, n. 4, suggests that it was given to Philippe de Dreux because of his martial character.*

[65] R and L add here "...traita, et Jehans Crisostomus en choisi en les (L - eles) natures des bestes et des oisiaus..."

[66] Turbulent seems a fitting epithet for this bishop who was twice a crusader, a prisoner in Rouen of Richard of England, a participant in the fighting against the Albigensians, and a lively figure at the battle of Bouvines. This information is taken from André du Chesne, *Histoire généalogique de la maison royale de Dreux* (Paris, 1631), pp. 33-45.

[67] Some years before he wrote the well documented article on bestiaries for the *Histoire littéraire de la France* where he expressed the above conclusions concerning identities, Paul Meyer in *Notices et Extraits des Manuscrits de la Bibliothèque Nationale et autres bibliothèques,* XXXIII, 1 (1890), pp. 10-11, had suggested that the recipient of the *Mappemonde* was Robert d'Artois, brother of Saint Louis, killed at Mansourah in 1250. However he did concede that this Count Robert could be one of the counts of Dreux, who bore this name in the first half of the thirteenth century. The evidence which perhaps led Meyer to settle more firmly on Robert II de Dreux is that another work of Pierre de Beauvais, *La Translation et miracles de saint Jacques,* dated 1212 and written in Beauvais, bears this inscription (L, f. 42): "Et Pierres par le commandement la contesse Yollent mist en romanz cest livre." This Yolland is probably the daughter of Raoul I de Coucy whom Robert II de Dreux married in 1184. Du Chesne, *op. cit.,* p. 45.

Several questions now arise. Why are there two versions? Which did Pierre write first? Where did he find the quantity of additional material? If the names of the patrons of Pierre were attached only to one version or the other, it could be supposed that *le grant bestiaire* had been written for Philippe de Dreux and the short form for his brother. Because of the spirited character of the Bishop of Beauvais it might be supposed that a more "exotic" type of bestiary would find favor with him, and that it is for this reason that Pierre "stuffed" a traditional bestiary with strange bits on birds and beasts. Unfortunately, there is no evidence whatsoever to support this hypothesis.

The editor of the only text in existence, Cahier, presumed that the long version was the original work, and that the short version had been consciously reduced in size.[68] Later, Lauchert decided in favor of the primacy of the shorter form.[69] This is the opinion expressed also by Paul Meyer[70] and repeated by Faral.[71] Since the relationship between R, L, and Ma and certain Latin manuscripts is on the whole too close to be the result of the amputation of extraneous material, this writer concurs with the above in thinking that Pierre's original composition was the short bestiary. His Latin source was a manuscript belonging to the B-Is version with the not wholly unusual addition of chapters on the Wolf and Dog.[72]

There remains the complex problem of sources for Pierre de Beauvais' long bestiary. It must be admitted immediately that no solution — not even a partial one — has been found as to the origin of the added descriptions, and that the suggestions offered here are of a very tentative nature.

Since Pierre introduces most chapters with the phrase "Phisiologes nos dist", the name *Physiologus* as an indication of a source

---

[68] Cahier, *op. cit.*, II, p. 97.

[69] Lauchert, *op. cit.*, pp. 137-38.

[70] Meyer, *op. cit.*, p. 386.

[71] Edmond Faral, "La queue de poisson des sirènes", *Romania*, LXXIV (1953), 489.

[72] It is pertinent to recall that Bodl., Douce 167 ends with a notice on the Wolf, and that B.M., Stowe 1067, although the chapters on the Wolf and the Dog are together in the middle of the manuscript, contains a text quite like Pierre's translation. Of greater significance than these two similarities, however, is the existence of an unillustrated fifteenth century Latin manuscript, Tours 312, f. 303-326v., which has exactly the same order of contents as the French B.N. 834, even to the rather uncommon final chapters (uncommon in that the Pearl [Margarita] is omitted): Amos, Adamas, Lupus, Canis.

has no value. Paul Meyer states that the extra chapters might
already have been interpolated in the Latin *Physiologus* from which
Pierre drew,[73] a view accepted by Faral.[74] It is true that an enlarged
bestiary of the Second Family (as represented by the complete Book
III of Pseudo-Hugo) could account for some of the supplementary
material in the long bestiary, in a very abbreviated form, but certainly not for all.

At present no single work has been found which could have
served as a source for all the additions.[75] There are, however, one
or two Latin works written prior to Pierre's bestiary that do contain
items similar to the French, though whether Pierre extracted directly
from these books or used them through an intermediary cannot be
ascertained. Similarities have been noticed between the accounts of
the *Harpia* (XLI) and of the *Musca et Aranea* (XVIII) in Odo of
Cheriton and those on the Harpy and on the Spider and Fly
contained in Pierre de Beauvais. Alexander Neckam's *De naturis
rerum* might have been used in composing the chapters on Argus
(Neckam i.39), the Barnacle Goose (i.48), the serpent Tiris (ii.108),
the Basilisk (i.75) and possibly for part of the section on the
Elements and Senses (ii.152). Still there remain numerous chapters
unaccounted for.

Relatively unexplored and totally unexplained thus far is the
relationship between certain descriptions in Pierre de Beauvais' long
bestiary and corresponding notices in the well known *Image du
monde* of Gauthier, or Gossouin, de Metz, written in 1246.[76] The
strange name of the animal resembling a flying squirrel in the
bestiary, *muscaliet,* appears again in the *Image du monde*; there
is a serpent called *tygris* in each compilation; the Echineis, Parrot,
Barnacle Goose, and Phoenix, among other accounts, are described
in strikingly similar terms in both works, though it is possible that
each author was using related sources. The mingling of the characteristics of the Leucrota and the Yale in a beast called the *centicore*

[73] Meyer, "Les Bestiaires", p. 386.
[74] Faral, *op. cit.,* p. 489.
[75] Celebrated Latin compilations such as those by Jacques de Vitry,
Thomas of Cantimpré, Vincent of Beauvais, and Bartholomeus Anglicus
offer no help in providing clues as to where Pierre might have found his
supplementary chapters.
[76] Oliver H. Prior (ed.), *L'image du monde de maître Gossouin* (Lausanne, 1913). Of possible import is the fact that both the Arsenal and
Montpellier manuscripts of Pierre's bestiary also contain the *Image du
monde.*

in both French compositions has been commented upon by Druce who concludes: "...both the writer of the French bestiary and Gautier de Metz obtained their information from some older bestiary, which had described these two animals the one after the other, but it is difficult to say at what time the confusion took place."[77] Surely the clarification of this problem would throw light on the still unsolved puzzles of Pierre's works.

With the composing of Pierre's short bestiary the ancient *Physiologus* tradition ends in France. The bestiary which in a large measure is derived from Pierre's long version, the *Bestiaire d'Amour* of Richard de Fournival, is, as already noted, totally secular in nature although many of the animal descriptions continue the early traits.

From this survey of the Latin versions and of the various French translations of the *Physiologus*, let us now turn our attention to the appearance of the manuscripts themselves, and more particularly to the illustrations which almost inevitably occur in the French manuscripts and which are very common in the Latin.

---

[77] Georges C. Druce, "Notes on the History of the Heraldic Jall or Yale", *Archaeological Journal*, LXVIII (1911), 185. See also the section on the Yale in Chapter V of the present study.

CHAPTER IV

ILLUSTRATED BESTIARIES

Many Latin bestiary manuscripts were illustrated, though it is difficult to say whether or not they were in the majority,[1] and it is quite rare to find a French manuscript without illustrations. England seems particularly to have favored Latin manuscripts with pictures, and illustrated Latin examples were by no means unusual in France. Indeed, the illustrations might be one of the factors contributing to the popularity of this type of work, for, as the authority on bestiaries in England, M. R. James, said, the bestiary ranked with the Psalter and the Apocalypse as one of the leading picture books of twelfth and thirteenth century England.[2] From the carefully and ornately executed miniatures of the more sumptuous Latin manuscripts to those carelessly drawn and sometimes unfinished in the vernacular versions, the idea of illustrating the obvious traits of the animals and birds occurred to many artists. The subject in most cases lent itself well to visual portrayal, like the capture of the Unicorn or the Phoenix on its funeral pyre; on the other hand much ingenuity was demanded in the depiction of certain scenes which were more difficult to translate into immediately comprehensible pictures, such as that of the Onager braying at the equinox (Pl. VII, Fig. 2) or the Fire Stones which are male and female by nature. These are often portrayed in the form of a man and a woman holding stones and surrounded by flames. It is evident

---

[1] None of the manuscripts of the version ascribed to Theobaldus that have been seen has contained illustrations, nor, strange to say, has any of the simple B version, but it is hard to believe that in some library in Europe there is not at present an illustrated B manuscript which somehow has been overlooked.

[2] James, *op. cit.*, p. 1.

that the treatment of the fabulous or rare animals offered the artist the greatest liberty for his imagination, that of the domestic animals the least, while the depiction of the birds was often perfunctory and undistinguished. In all instances, however, the pictures are valuable not only as examples of the developement of mediaeval illustration in succeeding periods, but their aid in revealing common or curious interpretations of the text is immense. This self-evident observation makes it all the more surprising that relatively little study has been devoted to the illustrations.[3]

It is assumed that the earliest Latin translations of the *Physiologus*, which, as noted before, were probably first made around the fourth century, took over the pictures of the Greek manuscripts as well as the text. Unfortunately, what is probably the oldest Greek manuscript of the *Physiologus* is of a much later date and contains no miniatures (Morgan 397, late tenth century),[4] and the only early Greek manuscript that has been described in any detail is the Smyrna Codex of about 1100.[5] Its miniatures portray not only the characteristics of the animals but also the religious allegories exposed in the text. Because this manuscript contained miniatures illustrating the *Christian Typography* of the sixth century author Cosmas Indicopleustes, which Stryzgowski thought was both written and illustrated at Mount Sinai, this art-historian suggested an archetype of Syro-Egyptian creation dating possibly from the sixth century for this Greek *Physiologus*.[6] It appears though that the Cosmas miniatures are not of Sinaitic origin, but rather of Alexandrian,[7] and

---

[3] The only scholars who have treated the illustrations at any length, and more often than not their investigations have been focused on individual manuscripts, are Strzygowski, M. R. James, G. Druce, H. Woodruff, D. Tselos, S. Ives and H. Lehmann-Haupt, and H. Menhardt, all of whose works are referred to in the course of this study.

[4] This manuscript was pointed out by B. E. Perry in his review of Sbordone's edition of the Greek *Physiologus* which appeared in the *American Journal of Philology*, LVIII (1937), 495.

[5] Josef Strzygowski, "Der Bilderkreis des griechischen Physiologus", *Byzantinisches Archiv*, Heft 2 (1899), pp. 1-130. This manuscript, B.8, was destroyed in 1922 when the Library of the Evangelical School was burned by the Turks. It is regrettable that Strzygowski did not publish more plates of illustrations rather than present the majority of them by written description only.✝

[6] *Ibid.*, p. 99.

[7] For a recent affirmation that Cosmas wrote in Egypt, see Milton V. Anastos, "The Alexandrian Origin of the *Christian Topography* of Cosmas Indicopleustes", *Dumbarton Oaks Papers*, No. 3 (Cambridge: Harvard University Press, 1946), pp. 73-80.

thus another element seems to point to Alexandria as the probable place of origin of the *Physiologus*.

Of the approximate forty-eight miniatures in the Smyrna manuscript which portray animals, most af the subjects illustrated, except for the marked difference in style, are very similar to those found in the Latin manuscripts. It should be remembered, however, in this connection that the very nature of the material gives the artist little occasion for marked deviation. Allegorical personifications, not to be found in the Latin illustrations, appear in the scene of the Eagle where a nimbed figure represents the nymph of the fountain, and in that of the Owl where this bird sits on the hand of a woman symbolizing Night. Among the more interesting illustrations, when compared with their Latin counterparts, are those of the Siren, a woman whose lower extremities are those of a full-feathered cock; the Unicorn, a large, cloven-hoofed animal whose long horn has a reverse curve, and which stands with raised foreleg before a seated woman; and the Phoenix, which is pictured both walking on the ground and in flames atop a column while it gazes at the sun. Beside the pillar stands a nimbed priest. Scenes that have not been found in any Latin manuscripts (because, except in rare cases, the material in the Greek text was not continued in Latin) are those of the Ant-Lion with the forepart of a lion and the hind part of an ant, the Ichneumon springing against an erect snake, the Frog in a pond, the Vulture sitting on a stone, and for the chapter on Amos and the fig tree a man in a tree collecting figs. Illustrated manuscripts of the Greek *Physiologus* apparently were less popular than the Latin versions, and it is therefore to the latter that one must turn for an abundance of illustrations.

The oldest Latin illustrated manuscript is the one already referred to as C, Bern 318, of the ninth century. It will be recalled that its chapters, though not so numerous, follow the order and content of the early Greek and Ethiopic versions. The miniatures depict only the characteristics of the animals except for the first one, which portrays Jacob blessing the lion of Judah.[8] Stylistically the illustrations belong to the school of Rheims, the most distinguished of Carolingian artistic centers. In her study of the Bern *Physiologus* Miss Woodruff emphasized what she considered as

---

[8] Photographs of all the miniatures are reproduced in Miss Woodruff's article, "The Physiologus of Bern", *Art Bulletin*, XII (1930), 226-53, but in some the paint has peeled off, a condition which occasionally makes an adequate interpretation of the subject, based on these photographs, difficult.

traces of the illusionism which existed in early Alexandrian painting. Finding certain stylistic similarities between this *Physiologus* manuscript and the fourth century Vatican Virgil and the Ambrosian Iliad, among other manuscripts, and since the latter has been connected with Alexandrian illumination,[9] Miss Woodruff implied that the model for the Bern manuscript had its origin in Alexandria in the fourth century. Recently, however, the linking of the Bern *Physiologus* with the early manuscripts has been questioned, and a closer resemblance has been noted and studied.[10] Professor Tselos has indicated striking similarities in both human and animal figures between the Bern *Physiologus* and the well known Utrecht Psalter.[11] Because of its obvious relationship to the Utrecht Psalter and because the Bern *Physiologus* repeats a section from Isidore of Seville's *Etymologiae*,[12] which must have been incorporated in the model for the Bern manuscript after the early seventh century, Professor Tselos cannot assign to the archetype of this manuscript as early a date as did Miss Woodruff. Instead, he believes "that the model of the Physiologus was made in the same Greco-Italian center which made the model or the Utrecht and Troyes Psalter sometime in the seventh or eighth century."[13]

Notable illustrations in the Bern *Physiologus* are the portrayal of the two Vipers as having the upper part of a man and of a woman, each with a serpent's tail (Pl. IX, Fig. 3), the Siren whose lower extremities are those of a serpent — an interpretation which is contrary to the text — Jacob blessing the Lion of Judah, and the unique miniatures of the use of the Agate in finding Pearls and of the Indian Stone curing dropsy.

Next in age to the Bern manuscript is Brussels, Bibliothèque Royale 10074, of the tenth century. This manuscript, which also

---

[9] *Ibid.*, p. 238.

[10] Although Miss Woodruff briefly indicated certain similarities between the Bern *Physiologus* and the Utrecht Psalter, the subject is fully treated by Dimitri Tselos in "A Greco-Italian School of Illuminators and Fresco Painters: Its Relation to the Principal Reims Manuscripts and to the Greek Frescoes in Rome and *Castelseprio*", *Art Bulletin* XXXVIII (1956), 1-30.

[11] Especially convincing is the depiction of the Unicorn, which in the Bern manuscript, following the text, is portrayed as a goat with its horn curving back (Pl. IX, Fig. 2 a), the whole forming a unique conception. The Utrecht Unicorn is identical.

[12] This section is on the Horse (Caballus) and exists only in the Bern *Physiologus* among early manuscripts.

[13] Tselos, *op. cit.*, p. 10.

includes illustrations in the same style of the *Psychomachia* of Prudentius, contains no Alexandrian illusionism, but is a Carolingian reworking of an antique archetype.[14] In the drawings there is a marked tendency, of which this is the only known example among Latin manuscripts, to introduce pictorial references to the allegory. For example, in the second illustration of the Caladrius the figure of Christ appears, the tempting of Eve as Adam stands by is under the drawing of a figure holding two rings representing the Fire Stones, and the Crucifixion is depicted below the Pelican shedding its blood. Legends, at times rather lengthy, usually explain the scene portrayed. Some of the uncommon details seen in these illustrations are the capturing of the Caladrius with a net, the priest's discovering the Phoenix upon the altar, and the Owl perched on King David's palace. Unfortunately, fewer than half of the carefully executed drawings were completed in this manuscript.

Although there are doubtless illustrated manuscripts of the *Physiologus* dating from the eleventh century, none has so far been signaled. In general, the early twelfth century manuscripts contain pen drawings that still retained certain Carolingean features. Usually added after the text had been copied, the problem of adapting the drawing to the spaces which were left was one which a poor artist cannot be said to have solved successfully. Towards the end of the century, when the *Physiologus* was doubled in size by the numerous additions from Isidore, the illustrations themselves became far more elaborate. De luxe editions appeared, almost exclusively in England, with carefully painted miniatures correctly placed on the page, framed, and highlighted with gold — interesting compositions but often lacking the animation of the earlier drawings.[15] It is not unusual for the number of miniatures to be increased from around forty to over a hundred; Bodl. 764, for example, contains some one hundred and thirty-five illustrations. Characteristic of the best of these manuscripts is the arbitrary coloring, the love of symmetrical patterns, and the fact that artistic considerations take pre-

---

[14] See Helen Woodruff, "Illustrated Manuscripts of Prudentius", *Art Studies*, VII (1929), 48. The remark in the text applies specifically to the *Psychomachia* illustrations, but might be presumed to describe also the evolution of the *Physiologus* drawings. For illustrations of this manuscript see Chapter II, n. 13.

[15] Among the finest Latin illustrated bestiaries are the following: Leningrad Qu.V.I, Aberdeen 24, Ashmole 1511, Harley 4751, Bodl. 764, Royal 12 F. xiii and the Dyson Perrins Bestiary. Very modest examples are Valenciennes 101 and B.N., lat. 14429.

cedence over natural observations. Although it is probable that there were numerous model books in existence, such as the Hofer Bestiary,[16] it is unusual, though not unheard of, to find the illustrations of one manuscript copied in another.[17]

When we turn to bestiaries in French, we realize how exceptional it is to find an unillustrated manuscript, although there are cases, as in the Latin manuscripts, where the spaces left for the miniatures have remained empty. It is at once apparent also that in the majority of instances the illustrations in the vernacular versions are inferior in both conception and execution to those in Latin.[18] This is doubtless in large part due to the audience for whom the work was destined — a less wealthy, and evidently rather often a less discriminating group, than the one to whom the finer Latin versions were addressed. One such example, though admittedly an extreme case, is seen in the portrayal of the Aspidochelone (Pl. II. Figs. 1 a and b) where the first minutely detailed picture is a full-page painting in a Latin manuscript, and the second is a small, crudely executed drawing in a manuscript of Philippe de Thaon. The latter manuscript (Oxford, Merton Coll. 249) repeatedly uses the same beast to represent animals as different, to our knowledge at least,

---

[16] Ives and Lehmann-Haupt, *op. cit.* See p. 31, note 33.

[17] The following illustrated manuscripts whose relationships have not heretofore been noted appear either to be copies one of the other or to have been created by the same artist or in the same atelier. Since these resemblances have been seen only on microfilms, the conclusions must be tentative and should not be totally accepted until the manuscripts themselves are consulted. The outlines of many of the drawings in B.M., Add. 11283, early twelfth century, are pricked, and several, though not all, of the illustrations appear to have their counterpart in Brussels, Bibl. Roy. 8340 of the fourteenth century. Numerous careless drawings found in the unique Fourth Family manuscript, Camb., Univ. Lib. Gg.6.5 of the fifteenth century, look like poor copies of the rather heavy pictures in Copenhagen, Gl. Kgl. 1633 4°. Finally, although the script belongs to different periods, the framed pictures with their lengthy legends found in B.N., lat. 3630, fourteenth century, appear related to the imaginative illustrations in B.M., Harl. 3244, early thirteenth; details and scenes sometimes make one think that the difference was intentional — the miniaturist was trying not to repeat himself — but the style looks remarkably similar in certain instances.

[18] This statement does not apply to some of the manuscripts of Guillaume le Clerc's bestiary which occasionally are of excellent quality though still on a reduced scale in comparison with some of the more magnificent Latin manuscripts. Among the best French examples can be listed B. N., fr. 14969, 14964, 24428, and the small, delicate miniatures within the capital letters in Camb., Fitzwilliam Museum J.20.

as the Crocodile and the Elephant. The artist of the Queen Mary's Psalter, though skilled in producing spirited drawings, is also guilty of this repetition.

There are few changes other than the minor ones that would result from the general simplification characteristic of the French miniatures in comparison with the Latin. It seems inevitable that the drawings in some of the French manuscripts originally had as models the corresponding Latin drawings; [19] this hypothesis seems particularly applicable to the Copenhagen manuscript of Philippe de Thaon. As yet it has not been ascertained whether Latin models were used for the miniatures of some of the additional bestiary chapters found in Pierre de Beauvais' long *Bestiaire*. Unusual for both Latin and French versions are the two manuscripts of Guillaume le Clerc (B.N., MSS. fr. 14969 and 24428) which in addition to the customary animal scenes depict the moralization. A knowledge of the text is necessary for an adequate interpretation of most of these illustrations. Occasionally the content, though not the style, will hark back to a much earlier period, like the priest kneeling

---

[19] Influence flowing in the opposite direction, from the French miniatures to the Latin, might explain certain illustrations in two Latin manuscripts to which James attributes a non-English origin, Perrins 26 and Sion College L $\frac{40.2}{L\ 28}$ (James, *op. cit.*, p. 11). In the latter the Hydrus-Hydra confusion, found often in the illustrations of the *Bestiaire d'Amour* of Richard de Fournival, appears with a multi-headed creature protruding from the side of a bear-like Crocodile, and in both manuscripts the Salamander is a winged beast standing in flames, and so it appears on some manuscripts of both Guillaume le Clerc and Pierre de Beauvais. It is also possible, of course, that these French peculiarities could have arisen first in the Latin manuscripts. Stronger reasons exist for claiming French influence, though textual rather than pictorial, on the marginal bestiary drawings in B. M., Royal 2 B. vii. In his introduction to the facsimile reproduction of this manuscript, Sir George Warner (*Queen Mary's Psalter*, London: British Museum, 1912, p. 33) says he believes that the artist used as his source the *Bestiaire* of Guillaume le Clerc in a fuller form than is now known, or that he supplemented Guillaume from some other source. I am inclined to think that the main source for all of the drawings which follow for the most part the traditional order of B, with the addition at the end of nine extra animals, is the long version of Pierre de Beauvais. This conclusion is based, among other reasons, on the illustration of the Asp on f. 125v., which depicts the beast lulled to sleep by instruments as two men approach a tree as if to steal from it. This is without doubt the "arbre dont li baumes dégoute" of which Pierre de Beauvais speaks (Cahier, Vol. II, p. 149), and of which there is no mention either in Guillaume le Clerc or in any Latin bestiary.

before the burning Phoenix on an altar, seen in the fourteenth century manuscript B.N., 14969 of the same Guillaume. One would never study the illustrations in French bestiaries, any more than in the Latin versions, for evidence of the artist's awareness of the world of nature about him; a close visual interpretation of the text as he understood it or as his model had presented it was the object of the illustrator.

CHAPTER V

GENERAL ANALYSIS OF THE PRINCIPAL SUBJECTS TREATED IN LATIN AND FRENCH BESTIARIES

Previously there has been no single compilation in English in which were listed the characteristics of the birds, beasts, and stones described in the Latin and French bestiaries. In this chapter under alphabetically arranged subjects these traits are presented in their approximate chronological order,[1] except for Pierre de Beauvais' non-bestiary material, which has been relegated to the Appendix. The common Latin and French names are given as well as any conspicuous variations in spelling. Words enclosed in single quotation marks in Biblical citations indicate a pre - Vulgate reading as given by Professor Carmody in his edition of B and of Y. When it exists, the description begins with a summary of the earliest Latin accounts, Y and B, with any important differences existing in A, C, Theobaldus, or the *Dicta Chrysostomi*. This is then followed by the expanded versions such as Royal 2 C. xii, the *Aviarium* (Book I), or Books II and III of the *De bestiis et aliis rebus*, referred to here as H because of its traditional, though erroneous, attribution to Hugo of St. Victor. When H has not provided the fullest account, additional information has been taken from the more complete version in Cambridge, University Library Ii.4.26. Isidore's copious contributions are all noted. As for analagous passages in antiquity and in the Church Fathers, the examples given in this chapter —with the general exception of references to Aristotle's *Historia animalium* and Pliny's

---

[1] The question repeatedly arose of what English equivalent to give to an animal or bird which was already the object of confusion among early writers. This is the case, to cite only one instance, of the bird which was described under the name of *Erodius* and *Fulica* and which partook of traits assigned to both the Heron and the Coot. In attempting to resolve these problems, consistency and logic have been the aims in establishing the nomenclature used.

Naturalis historia— are far from complete, and are presented only to show the age and currency of many of the tales.[2] The significant changes, though few in number, in either additions or omissions occurring in the French bestiaries are noted, but no attempt has been made to gather other mediaeval accounts which are similar in nature.[3] In conclusion, a short iconographical description is given of each subject illustrated wherein marked deviations from the typical are indicated.

The allegorical interpretations, although perhaps the *raison d'être* of the *Physiologus*, have intentionally been treated very briefly. A summary has been made only of those chapters contained in the B version since this is the basic Latin text from which most others grew or were translated. Usually the allegory remained fundamentally unchanged when incorporated in the enlarged bestiary or when translated into French; however, Philippe de Thaon and Guillaume le Clerc often emphasized different parts of the "lesson", and the latter tended to embroider long moralizing passages on the rather bare original framework.

While it is realized that each topic has not been exhaustively investigated, and that many questions have been left unsolved, it is hoped that a certain insight has been provided into the imaginative reasoning of mediaeval man, who continually sought a logical explanation of the unknown or little known, whether it be in the boundless realm of metaphysics or in the more restricted question of how the fallen elephant arises.

For brevity the following abbreviations are used in this chapter:

Latin:

    A - Brussels 10074, X cent. (Cahier)
    B - Bern 233, VIII cent. (Carmody)
    B-Is - B.M., Royal 2 C. xii, XII cent. (Mann)
    C - Bern 318, IX cent. (Cahier)
    Y - Munich 19417, IX cent.
         " 14388, IX-X cent. } (Carmody)
        Bern 611, VIII cent.
    TH - Theobaldus, XI cent. (?) (R. Morris)

---

[2] An extensive list of sources, analogies, and derivative passages recorded by authors of antiquity and by early Christian writers exists in Sbordone, pp. 1-145.

[3] This aspect, with some reference made to classical sources, is treated by Max Goldstaub and Richard Wendriner, *Ein Tosco-Venezianischer Bestiarius* (Halle, 1892), pp. 255-438.

DC - *Dicta Chrysostomi*, XI cent. (Wilhelm)

H - Pseudo-Hugo of St. Victor, Books II and III of *De bestiis et aliis rebus*, XII cent. (?) (Migne)

CUL - Cambridge, University Library Ii.4.26, XII cent. (James)

French:

PT - Philippe de Thaon, *ca.* 1125. (Walberg)

G - Gervaise, early XIII cent. (?) (Meyer)

GC - Guillaume le Clerc, *ca.* 1210. (Reinsch)

PB - Pierre de Beauvais, before 1218. (Cahier)

AMOS AND THE FIG TREE.

psycomora, sycaminus, Amos propheta; Amon li prophetes.

A unique Latin account, entitled *De sycomora*, exists in Y (28), and begins: "Non sum propheta 'neque' filius prophete, sed 'caprarius' vellicans sycamina" (Amos 7:14). There follows a brief description telling of gnats which live in the figs in darkness until the fruit is opened and they come forth to light.

When transferred to the group of manuscripts known as B and B-Is, this chapter (34) is entitled *Amos propheta* and it omits the fabulous tale of the insects. In reality it contains only the lengthy moralization which, in its essential, says that just as Amos was a goatherd, Christ became the pastor of sinful mankind.[4] In this form the chapter was translated into only one French bestiary, that of PB (IV, 63).

Biblical commentators, though alluding to the splitting of the figs,[5] make no mention of the insects dwelling within, and Sbordone suggests that this account might have been combined by the author

---

[4] The second sentence of a typical B-Is manuscript, Bodl. 602, f. 32, after the quotation from Amos, reads "Et distincto more salvator per prophetam de se dicit...", which M.R. James believes to be a corruption of "et districtor morae". See James, *op. cit.,* p. 8.

[5] For a notice on the preferable translation of the word *vellicans* as "tending the fruit" rather than "piercing"—which refers to an ancient tradition of puncturing the fruit to improve its flavor— consult the note on Amos 7:14 in William A. Harper, *A critical and exegetical commentary on Amos and Hosea,* Vol. 20 of the *International Critical Commentary* (New York, 1915), p. 172, and the article on "Sycomore" in James Hastings, *Dictionary of the Bible* (New York, 1908-9), IV, 634-35. The fig-piercing and the insects dwelling therein are discussed in Konrad Burdach, *Der Gral,* Vol. 14 of *Forschungen zur Kirchen-und Geistesgeschichte* (Stuttgart, 1938), 34-40.

of the *Physiologus* with that of Aristotle (v 557b 26) on the fig-wasp which lives in the fruit of the wild fig.[6]

The Latin illustrations shows Amos seated, leaning on his staff, as his goats stand against trees and nibble. In Arsenal 3516, f. 210 (PB) only a haloed seated figure is drawn.

AMPHISBAENA.

amphisbaena, amfivena.

H's chapter (iii.44), a copy of Isidore's (xii.4.20), begins with an explanation of this reptile's name, so called because it has two heads, one in place and the other at its tail. This enables it to run in either direction by a revolving motion of the body.[7] It is the only serpent to expose itself to the cold, and it goes first before all. A quotation is given from Lucan's *Pharsalia* (ix.719): "Et gravis in geminum vergens caput amphisbaena", "the fell amphisbaena, that moves toward each of its two heads". In conclusion, its eyes are said to shine like lamps.

Pliny (viii.23.35) explains the amphisbaena's two heads "as though one mouth were too little for the discharge of all its venom". The origin of the shining eyes is not known. Today the name amphisbaena is applied to a legless lizard which is able to progress either forward or backward.[8]

One of the more common representations of the amphisbaena is that of a horned, winged, two-footed serpent, making a circle of its body and biting its tail, although it is also shown more prosaically as a two-headed worm (Pl. I, Fig. 1 a and b).

ANT, ANT-LION, GOLD-DIGGING ANTS.

formica, mirmicoleon, formicaleon; furmi, formicaleun.

Y contains both a chapter on the ant and on the ant-lion; to a combination of these H (ii.29) adds an account of the gold-digging ants of Ethiopia. The result is altogether a strange mixture.

---

[6] Sbordone, p. 144.

[7] From the Greek ἀμφίσβαινα, ἀμφίς, 'both ways', 'around' and βαίνειν 'to go'.

[8] For a detailed study on the mythological and the actual amphisbaena see George C. Druce, "The Amphisbaena and its Connexions in Ecclesiastical Art and Architecture", *Archaeological Journal*, LXVII (1910), 285-317.

Y (14) and B (11) begin with a quotation from Proverbs: "Vade ad formicam, o piger, et 'imitare' vias eius" (Prov. 6:6). They continue by saying that the ant has three natures: it walks in order carrying grain in its mouth, unmolested by other ants who have no grain; it divides each grain in two lest the rain cause it to germinate and the ant perish from hunger in the winter; it goes into the fields at harvest time and, climbing the ear of grain, distinguishes wheat from barley by its odor, and refuses the latter because it is food for cattle. CUL adds that ants take advantage of good weather to dry their grain.

In B's moralization, which serves as a general basis for the other versions, the improvident ants are equated with the Five Foolish Virgins. The necessary splitting of the grain represents the separation that must be made in the Old Testament between the literal and the spiritual meaning in order to prepare for the Day of Judgment (winter). The spiritual sense must transcend the literal, though the Jews failed in this. The barley which the ants flee is heretical doctrine (here B and PT list the names of certain heretics). PT expands the section on the Wise Virgins, who for him represent the five senses.

Entitled *De mirmicoleon,* Y (33) begins: "In Iob Elefas Temaneorum rex dixit de mirmicoleon: Periit eo quod no habeat escam" (Job. 4:11). The ant-lion is described thus: its father has the face of a lion and eats flesh; its mother has the face of an ant and eats plants. When the *mirmicoleonta* is born, it dies because of its two natures, unable to eat either flesh or plants because of its lion's face and its fore and hind quarters of an ant.

H and B-Is (11), naming the *Etymologia* (xii.3.9) as their source, say that the *formica* is so called because it carries morsels of grain ("ferat micas farris"). They continue by describing the *formicaleon,* thus named either because it is the lion of ants, or equally an ant and a lion. It is a small animal so hostile to ants that it hides in the dust and kills the ants bearing grain. Thus it is called ant and lion, because as the lion is stronger than other beasts, so this animal is stronger than other ants.

The same chapter of H and B-Is includes the tale of the gold-digging ants said to exist in Ethiopia. These ants, the size of dogs, dig up golden sand with their feet and guard it. Should it be carried off, they pursue and kill the thieves. To steal the gold, people capture mares with foals. Starving the mares for three days, they tie the foals at the edge of the river flowing between them and the ants, while the mares with pack saddles are driven across the water to fields of green grass where they pasture. When the ants see the pack saddles, they bring the golden sand and hide it in them. Toward evening the mares hear their foals neighing from hunger, and so return to them laden with gold.

One must wonder about the origin of the disparate elements which make up the chapter on the ant and the ant-lion. Concerning the traits of the former, distantly related passages can be found in Pliny (xi.30.36) and Aelian (ii.25). Research on the ant-lion has been made by George Druce, whose findings are here summarized.[9] The inclusion of the *mirmicoleon* in the *Physiologus* is due to the appearance of this word in the above cited passage in the Septuagint version of Job. H's description of the ant-lion closely follows that in Isidore (xii.3.10), who in turn repeats what Gregory had written in his *Moralia* on Job (V 20.40).[10] The Hebrew for the animal spoken of is *lajisch,* an unusual word for "lion". In the Vulgate this is translated *tigris,* in the King James version "old lion", and only in the Septuagint does μυρμηκολέον appear. There was a belief in antiquity that an animal called μόρμηξ existed. In his description of Arabia, Strabo (xvi.4.15) mentions a country abounding in lions called ants.[11] According to Druce the eastern version of the ant-lion tale —that of its double nature from its ant and lion parentage— had its origin in the Greek *Physiologus,* and it was not until the time of Gregory that a cleavage took place between eastern and western interpretations.[12] James has pointed out that for some unknown reason the name *Mermecolion* was attached to the pearl-oyster in a few of the Latin bestiaries.[13]

The story of the gold-digging ants is found in Herodotus (iii.102-105) where the place of origin of these ants, which are smaller than dogs but bigger than foxes, is India. Strabo (xv.1.44), citing Nearchus and Megasthenes, speaks of the μυρμήκες χρυσωρύχοι in India, as does

---

[9] George C. Druce, "An Account of the μυρμηκολέων or Ant-Lion", *The Antiquaries Journal,* III (1923), 347-64.

[10] Migne, *Patr. Lat.,* LXXV, Col. 700.

[11] Cf. Aelian vii.47.

[12] Druce, *op. cit.,* p. 354. "...for while the eastern imagination and love of the picturesque nursed the idea that the ant-lion was composed of an ant and a lion, Gregory and the more sober western commentators adopted the view that the ant-lion was no more than a large ant which preyed on smaller ants. This passed into Isidore's etymology, which set the seal upon it for the future".

[13] James, *op. cit.,* p. 9. I do not think that the names of the pearl and of the ant-lion are related.

Aelian (iii.4; xvi.15). Solinus changes their abode to Ethiopia and it is there, citing Isidore, that the earliest French bestiary locates them.

PT (861-1104) and GC (871-1020), after enumerating the three traits of the ants, describe gold-hoarding ants who resemble dogs. PT mentions the heavy load in comparison to the size of its body carried by the ant, and the fact that if rain wets the ant's grain, it throws the grain out. They both conclude with a description of the *formicaleon*, the lion of ants, cunning and harmful. G and PB limit themselves to listing the three habits of the ant.

Of all the illustrations included in the bestiary, probably those of the ant are the least satisfactory in clearly indicating the trait to be depicted. The ant itself is sometimes a series of dots (Bern 318, f. 12v.), or again a multi-legged, bean-like figure in Morgan 832, f. 7v., where it is shown carrying kernels up a hill. Brussels 10074, f. 146 illustrates Solomon on his throne beside stalks of wheat which ants mount. Probably the most imaginative portrayal of ants is found in Oxford, Merton Coll. 249, f. 4v. (PT). Here the artist has used the center margin between the double columns of writing as the path up which the ants march by two's while near the top of the folio some turn to the right and walk between two lines of the text. The *formicaleon* mentioned by GC is shown as two dogs in B.N., fr. 1444, f. 245. In an especially fine manuscript of GC the gold-digging ants appear (Pl. I, Fig. 2). And in B.M., Royal 2 B. vii, f. 96 armed men combat the dog-like Ethiopian ants. However, apart from these, the illustrations are disappointing.

ANTELOPE.

autolops, antalops, antula, autula; aptalon, aptalops, antule.

This animal is so wild that no hunter can approach it. Its horns, which are saw-like, can cut tall trees, but when thirsty the antelope goes to drink in the Euphrates and there catches its horns in the abundant branches of a shrub which in Greek is called *herecine*. After the antelope's struggles do not free it, it cries out and is heard by a hunter who arrives and kills the animal (B 2; Y 2).

The antelope's two horns are the two Testaments by which man can cut himself free from vice (G says that the horns are abstinence and obedience). Man should especially beware of drunkenness, which might lead to lust and eventual death by the devil. GC admonishes clerks from playing in the thicket of worldliness, where pleasures kill body and soul.

With no Isidorean additions the story is repeated in H (ii.2) and in all the French bestiaries with no change in outline. PT (757-798) says that the antelope resembles a goat, and that its horns

> E si sunt endentees,
> Cum falcilles curvees, ... (765-766)

In one of the few similes found in the bestiaries, GC (239-278) describes how the antelope is caught:

> E ele est prise el roncerei
> Com un peisson en une rei. (266-267)

G (449-472) gives the source of the Euphrates as Paradise, and through some confusion, PB (II,116) believes that the Greek word *hericine* refers to a place which is full of small branches.

This animal is not related to the antelope that is known today, nor were the early compilers certain of its identity. The wide variety of spellings of the name of the antelope attests the lack of clarity as to what animal was being described, and some Greek and Latin manuscripts contain no name at all.[14] Outside the *Physiologus* both the name and the story are unknown in ancient literature, although not unsimilar accounts by two Arabic writers are given from a secondary source in Sbordone.[15] In the *Hexaemeron* attributed to Eustathius of Antioch (*ca.* 330) the word ἀνθόλοψ appears, and it has been suggested that this is a Coptic word in an already modified form.[16] Professor B. E. Perry thinks it not unlikely that the ῦδρωψ in the Greek *Physiologus* is a corruption of ὄρυξ.[17] The Oryx is described in Oppian's *Cynegetica* (ii.445 ff.) as a fierce animal which dwells in thickets. Its sharp horns become so firmly imbedded in the animal against which it charges that, unable to free itself, they both die.

Since the artist had no definite animal on which to model the antelope of the *Physiologus* story, it ranged from a horse-like figure with feathery horns in Brussels 10074 (Pl. I, Fig. 3) to one small

---

14 PW, p. 1091.
15 Sbordone, pp. 116-17.
16 PW, p. 1091.
17 *Ibid.,* p. 1092.

and dog-like with almost straight long horns in GC (Camb., Trin. Coll. 0.2.14, f. 34v.). Sometimes the essential feature of horns having *serrae figuram* was forgotten, and then again there are cases of great exaggeration. The scene of the antelope alone is portrayed as well as the more inclusive one of the animal's drinking while its horns are caught and a hunter pierces it, as in the early twelfth century Bodl. 602, f. 3 where the horns are unmistakably serrated. In spite of the specific expression in PT, the animal drawn in Oxford, Merton Coll. 249, f. 4 is hornless.

APE.

simia; singe.

The role of the ape in the Greek and earliest Latin *Physiologus* texts is mentioned in the chapter on the onager with which it was linked. There it is stated that at the equinox the ape urinates seven times (Y 25); it also has the appearance of the devil, having a head but no tail (B 21).[18]

H (ii.12) and B-Is (22) begin with one of Isidore's etymologies (xii.2.30-33): the *simia* is so called because in it the *similitudo*, "appearance", of human reason is sensed.[19] Apes rejoice at the new moon and grow sad when it wanes. There are five types of apes: the *cercopithecus* has a tail; the *sphinx* is rough haired and docile, forgetting its wildness; the *cynocephalus* is like a monkey with a long tail and a dog's face, whence its name; the *satyrus* has a pleasing face and lively gestures; and the *callitrix*'s appearance is wholly different from the others, since it has a pointed face, long beard, and wide tail. According to others *simia* or *simus* is a Greek name, for σίμος is the equivalent of *fimus*, "excrement", because the ape is a dirty, horrible beast with a flat and wrinkled nose. The female ape gives birth to twins: one it loves, and the other it hates. When hunted, the mother carries the loved one in front of her while the other must cling to her back. But as she tires, since she is running on two feet only, unwillingly she drops the preferred child, and thus the hated one is saved.

---

[18] The Egyptian origin and later Christian connection of these two traits have been studied by H.W. Janson, *Apes and Ape Lore in the Middle Ages and the Renaissance* (London, 1952), pp. 16 ff.

[19] Isidore, however, added "sed falsum est", which H omits.

The ape symbolizes the devil who had a head but no tail; that is, he had a beginning when he was in heaven, but because of his inner hypocrisy and deceit, he lost his head. PT differs in saying that the ape-devil mocks those who do evil, and will carry them in front of him to hell, while he leaves the good at his back with God.

The French versions remain close to the Latin.[20] In the few lines given by PT to the ape (1889-1899) it is noted that this animal imitates what it sees, makes fun of people, dirties itself when angry, and carries its favored offspring in front. GC (1927-1942; 1953-1964) calls it an ugly filthy beast which thinks evil, and though he states that there are more than three kinds, he only mentions those with a dog's head and the lunar-sensitive apes. G (361-366) merely repeats the early Latin description of its resemblance to the devil. The ironic import of the tale of the favorite offspring's being abandoned is perverted in PB (III, 230), where the exhausted mother leaves the less loved child.

The apparent similarity between the word *simia* and *similitudo* and the ape's actual resemblance to a human being early formed a conection in men's mind which was celebrated in a well known saying by Ennius quoted by Cicero in the *De natura deorum* (i.35.97), and cited throughout the Middle Ages: "Simia quam similis turpissima bestia nobis".[21] The division of apes into various types is found in Isidore, Solinus (27.56-60), and ultimately in Pliny (viii.54.80). The origin of the theme of the mother carrying her young can be traced back to the Aesopic fables of Greece.[22] As it is treated by Pliny, Horapollo (ii.66), and Oppian (*Cyn.* ii. 604 ff.) the mother, showing great affection for its child, embraces and often smothers it, but Avianus introduced the new element of the mother's being forced to drop the loved one. In a very abbreviated form this was adopted by Solinus and eventually by Isidore.

---

[20] An amusing addition to the account of the ape appears in Richard de Fournival's *Bestiaire d'Amour* (19,3) and because of its popularity deserves telling here. The wise hunter, knowing the monkey's penchant for imitation, puts on and takes off his boots where a monkey can see the action. Departing and hiding, the hunter leaves a boot, which the animal immediately tries on, but before it can be removed, the monkey is caught. The catching of the ape by shoes weighted with lead is found in Aelian (xvii.25); Pliny and Solinus mention boots and also the fact that hunters leave bird lime for the ape to rub in its eyes.

[21] Janson, p. 23, n. 9.

[22] *Ibid.,* pp. 31-32.

In the section on the Elements and Senses included in the Appendix to this study, it is noted that the monkey represents the sense of taste. The probable explanation for this choice is that the monkey was associated with the Fall of Man, and it is often pictured eating an apple.[23]

That many mediaeval artists had yet to see any member of the simian family is evident from the fanciful representations found in some manuscripts, where the scene most frequently portrayed is that of the huntsman pursuing the mother to whom one young clings while the other is carried. An unusual group is that including the satyr and the apple-eating ape in Bodl. 602, f. 18v. (Pl. I, fig. 4), or that in Bodl., Laud Misc. 247, f. 153v., where shaggy apes appear to be playing over the body of a recumbent man. The French illustrations are usually similar to the Latin, although in B. N., fr. 14964, f. 147 (GC) the ape, with two dogs nipping its heels, is fleeing on all fours with the young riding in a precarious fashion.

Some of the Latin bestiaries include other types of apes in separate drawings. The bearded satyr standing crosslegged and the mule-faced callitrix holding a cup appear in Bodl. 764, f. 17v., and horned satyrs with tails and cloven hoofs, one with an axe and shield, the other clutching a snake, are found in B. M., Harl. 3244, f. 41v.

A S P .

aspis; *aspis.*

The asp is often included in the same chapter with the Weasel (mustella). According to H (ii.30) and B-Is (27), which cite Isidore as their source (xii.4.12), the asp is so called because by its bite poison is discharged, for *as* or rather ἰός is the Greek word for "poison". H adds that others say that it means "I defend" because *aspiso* with that meaning comes from *aspis* (ἀσπίς), a "shield". The asp is said to avoid enchantment by songs intended to draw it forth from its cavern by pressing one ear to the ground and closing the other ear with its tail. Other poisonous serpents which are described in the same section with the asp will be mentioned at the end of this article.

Allegorically the asp represents the wealthy who press one ear to earthly desires (or, as PT says, they have one ear to the ground

---

[23] *Ibid.*, p. 240.

to amass wealth) and whose other ear is plugged with sin. GC contains a lengthy sermon on the evils of wealth, which he illustrates with a story of a rich man who cast his gold into the sea before it drowned him.

The deafness of the adder is spoken of in Psalm 58:5,6: "...they are like the deaf adder that stoppeth her ear; which will not hearken to the voice of charmers, charming never so wisely", and it is Saint Augustine in a sermon on the feast of Saint Stephen who is first credited with noting the ingenious device of plugging both ears at once.[24]

The French versions follow the Latin closely. PT (1615-1628, 1655-1679) even copies the etymology found in B-Is:

> As en griu venim est
> Dunt aspis numez est; (1655-1656)

He closes with a brief list of the noxious effect of certain unnamed serpents which are named and described in GC (2553-2588). A strange change takes place in G (1151-1168). The enchanter, instead of being a human, is another beast, who hates the asp and would kill it if possible.[25] An addition is made by PB (II, 148), where the function of the asp is to guard the tree from which balm drips. If a man wishes to procure this balm, he must put the asp to sleep by means of instruments.[26] Parallels to the asp's guarding the tree of balm are found in Herodotus (iii.107), where winged snakes watch over spice-bearing trees, and in Pausanius (xxviii.2-4), where vipers of Arabia protect the juice of the balsam trees.

---

[24] Migne, *Patr. Lat.*, XXXVIII, Col. 1432.

[25] A suggestion has been made by the present author that the Latin word *marsus*, meaning "enchanter", which appears in some Latin manuscripts has been misinterpreted by G to indicate some kind of beast. This is apparently what happened in the Icelandic *Physiologus*, where *marsus* appears in the text and the illustration depicts a four-footed animal. See F. McCulloch, "The Metamorphoses of the Asp", *Studies in Philology*, LVI (1959), 7-14. For the fragmented text and illustrations of the Icelandic *Physiologus*, see Halldor Hermannsson (ed.), "The Icelandic Physiologus", *Icelandica*, XXVII (1938).

[26] Still another version is found in Bruneto Latini's *Li Livres dou Trésor* (i.138), where the asp is said to carry a precious stone called a carbuncle which the enchanter hopes to obtain by saying certain words, but is unsuccessful because the asp deafens itself. This stone is the *dracontis* or dragon stone described by Pliny (xxxvii.10.57) and Solinus (30.16), which in two mediaeval lapidaries has been identified with the carbuncle. See Leo J. Henkin, "The Carbuncle in the Adder's Head", *MLN*, LVIII (1943), pp. 34-39.

Illustrations of the asp are interesting for their variety within a fixed theme. In the Latin bestiaries there is often a long-robed enchanter holding a book or scroll and standing before the semicircular serpent, which sometimes not very convincingly deafens itself. A slightly different scene appears in CUL (f. 48v.), where a man protected by a shield strikes at the winged asp's ear with a stick. A second illustration in Morgan 81, f. 83 shows a man holding a cloak before his face as a horned serpent with a golden stone embedded in its head gazes at him. In Queen Mary's Psalter (B.M., Royal 2 B. vii, f. 125 r. and v.) a quartet of musicians lulls the four-footed beast to sleep, while in a second drawing a portable organ and pipes are played as two men steal balm from a tree.

Other types of serpents are listed along with the asp in the early B-Is and H versions, although in the later enlarged Latin bestiary they are relegated to a separate section on Reptiles. It will be seen that there is an occasional confusion of names and traits.

Dipsas.—This serpent, as its name implies, causes death from thirst by its bite. The Latin name, as given by Isidore (xii. 4.13), is *situla*. The dipsas is so small that it is not seen when tread upon, and its poison kills before it is felt (H iii.49). The additional chapter in H is taken in part from Book IX of Lucan's *Pharsalia,* of which three lines are quoted telling of the death of the standard-bearer Aulus from the bite of the dipsas (11. 737-760). GC (2563-2566) also recounts the death by thirst. The dipsas is usually depicted in a conventional worm-like form with the occasional addition of wings or horns. However, in Bodl. 764, f. 98v. it appears as a small snake underneath a human foot.

Hypnalis.—Again from its name it could be deduced that this serpent kills by sleep, and it was from the hypnalis that Cleopatra received her death (H ii.30; Solinus 27.31). PT attributes her death to asps which she put to her breasts, where they drank so deeply of her milk that they sucked out her blood (1671-1679). B-Is and GC call this serpent *prialis*.

Haemorrhois.—After the bite of this serpent one sweats blood in such a way that life itself pours out through the veins, for blood is called *hema* ($\overline{\alpha}\overset{\imath}{\mu}\alpha$) in Greek (H. ii.30; Isidore xii.4.15; Solinus 27.32). Again it is in Lucan's *Pharsalia* (ix.805-814) that a description is given of the fatal effects of the haemorrhois. GC (2575-2583) says that the veins burst when this blood-colored snake bites.

Prester.—This serpent runs with its mouth open and steaming. Its bite causes such swelling that the victim dies because putrefaction sets in. Lucan (*Pharsalia* ix.789-804), from whom a line is quoted, describes a distended body untouched even by birds and wild beasts. PT (1665-1666) mentions those who swell.

Seps.—By the bite of this serpent the entire body is consumed. H again quotes from the *Pharsalia* (ix.723).

ASPIDOCHELONE.

aspidochelone, aspido de lone, aspido testudo, balene; *balain, lacovie.*[27]

B (24) begins with the statement that there is a sea monster called in Greek *aspidochelone* and in Latin *aspido testudo*. Because this large creature's back is covered with sand, sailors think it an island on which they alight. To cook their food they make a fire, but when the whale feels the heat, it submerges and drags the ship to the depths. When hungry, the whale opens its mouth and emits a pleasant odor. Small fish are thus attracted into the whale's mouth, which soon closes on them (Y 30). To the similar accounts of B-Is (25) and H (ii.36), CUL mentions Jonah and adds a section on the *balene*, both of which are taken from Isidore (xii.6.7-8). The Old French versions do not deviate from the early Latin.

According to the allegory the sailors are the incredulous who, ignoring the wiles of the devil, put their trust in him and sink with him to hell. The small fish are men of little faith who are destroyed by the lures of the devil, while those of great faith know his tricks and avoid him.

Albert S. Cook, who has written on the legend of the aspidochelone, or shield-turtle, has traced the germ of the *Physiologus* chapter to an account by Nearchus, admiral of Alexander the Great's fleet.[28] His experience with a disappearing island can be found in the *Indica* of Arrian (xxxi) and in Strabo's *Geography* (xv.2.13). A new element enters the story in an apocryphal letter of Alexander to Aristotle which appears in the romance of Pseudo-Callisthenes. Here it is reported that the occupants of a boat drowned when the

---

[27] For an explanation of the derivation *lacovie* from *Jacoines*, as the whale is called in the Old French *Voyage of St. Brendan*, see F. McCulloch, "Pierre de Beauvais' *Lacovie*", *MLN*, LXXI (1956), 100-1.

[28] Albert S. Cook (ed.), *The Old English Elene, Phoenix, and Physiologus* (Yale University Press, 1919), pp. lxiii-lxxxv. See also Cornelia C. Coulter, "The 'Great Fish' in Ancient and Medieval Story", *Transactions of the American Philological Society*, LVII (1926).

supposed island which they were approaching proved to be an animal and sank. Finally, the *Physiologus* tale is quite similar to that told in Hebrew by Rabbah bar Hana, a Babylonian rabbi of the late third century A.D. As for calling the sea monster a turtle, there are numerous and widely spread legends about the size and habits of turtles which might have led to the apparent confusion. No explanation has been given for the odor attracting fish, but it should be noted that the aspidochelone in the oldest Latin texts follows the chapter on the panther where a similar trait is attributed to that animal.

The mediaeval imagination's treatment of a relatively uncommon sea creature is evident in the depiction of the aspidochelone, where a big fish is the base to which were occasionally added such non-cetacean features as legs. For one interested in the appearance of mediaeval ships, there are numerous examples as well as a picturesque variety of camp-fire scenes. Often the two traits of the whale are combined in one picture as in Bodl. 602 (Pl. II, fig. 1 a). In Camb., Sidney Sussex Coll. 100, f. 42v. an unwary sea traveler dries his stockinged feet over the fire burning on the fish which, in the Leningrad Bestiary is provided with large tusks. Here also the sailors in the ship above the whale but not on its back busily raise the mast. The miniature depicting *la coines* in B.N., fr. nouv. acq. 13521, f. 27 of PB shows a strange beast with heads at the tips of its three tails, though the illustrations in other PB manuscripts are normal.

ASS.

asinus, asellus.

Copying Isidore (xii.1.38), H (iii.22) says that the name of the ass and its colt comes from *assidendo*, for they are "sat on" by men. This animal is slow and resists commands. Large, tall asses are Arcadians;[29] the smaller ones are more useful since they sustain hardships.

Usually a single well-drawn ass is portrayed; more rarely a genre scene appears such as that in Bodl. 764, f. 44, where the ass, with a sack on its back, stands at the door of a mill.

---

[29] Pliny (viii.45.68) speaks of asses from Arcadia as being the most valuable.

BASILISK.[30]

basiliscus, regulus, sibilus.

H (iii.41) combines passages from Isidore (xii.4.6-9) in a chapter which begins by explaining that the Greek word *basiliscus*, "little king", is *regulus* in Latin, for the basilisk is the king of serpents, which it kills by odor just as it kills men by its glance. No bird can fly before its sight unharmed, but is consumed by fire from its mouth. The weasel, however, can kill the basilisk. It is a half foot long and marked with white spots. Like scorpions the basilisk seeks dry places, and after coming to water it bites people, making them hydrophobic and lymphatic. The *sibilus*, which kills by hissing, is the same as the *regulus*.

Most of H's account comes from Pliny (viii.21.33; xxix.4.19), and from Solinus (27.51), who specifies that the white spot on the basilisk's head resembles a sort of diadem. This would give rise to the identification as a "little king". The sentence on this serpent and the scorpions is taken in a slightly changed form from Jerome, *Epistola LXIX ad Oceanum.*[31] The difficulties of translating the names of reptiles in the Old Testament are well illustrated by the indiscriminate use of adder, basilisk, and cockatrice by the translators.[32] This is mentioned because in the mingling of names might lie the explanation for the basilisk's birth from a cock's egg, as described in the Appendix.

The basilisk is drawn in numerous forms. Camb., Corpus Christi Coll. 53 (Pl. II, fig. 2) shows the crested cock with a snake's tail attacked by its enemy the weasel. B.M., Harl. 3244, f. 54v. has two illustrations: the one entitled *basiliscus* shows a crested snake darting its forked tongue at a man bearing a spear and protected by a shield; the other of the *regulus* depicts a man falling and another fleeing before the approach of two snakes, one of which is crested. In Bodl. 533, f. 22v. the *regulus* is depicted as a dog walking in water.

---

[30] PB's non-bestiary account of the basilisk is found in the Appendix. A lucid presentation of the basilisk is one of the many animals appearing in the attractive and scholarly study of P. Ansell Robin, *Animal Lore in English Literature* (London, 1932), pp. 86-91 and Appendix.

[31] Migne, *Patr. Lat.,* XXII, Col. 660.

[32] See the article "Serpent" in Hasting's *Dictionary of the Bible,* IV, 459-60.

BAT.

vespertilio.

H's account (iii.34) differs somewhat from that of Isidore's (xii.7.36), and begins by saying that this ignoble bird takes its name from *vespers*, "eventide". It is both a bird and a quadruped, differing from other birds by having teeth and by giving birth to living young instead of eggs. For flying it is supported by a membranous wing. Bats hang from high places like a cluster of grapes, and if one loosens its grip, they all do.

A ventral view of the bat is usually shown with a minimum of direct observation evident. A lively drawing is that in B.M., Royal 2 B. vii, f. 91v. and 92, where men brandishing branches drive off a bat.

BEAR.

ursus.

From Isidore (xii.2.22), H (iii.6) takes the etymology that since the bear forms its offspring with its mouth, *ore*, *ursus* resembles the word *orsus*, "beginning". Bears do not conceive like other quadrupeds, but with mutual embraces like humans. Winter arouses lust, but the male respects the pregnant. After thirty days the female gives birth to a small piece of white flesh which is eyeless. Gradually it is shaped by licking and is kept warm by being held to the chest, and life is breathed into it. Meanwhile no food is taken, and for fourteen days bears sink into such a deep slumber that even wounds do not arouse them. When they emerge, their eyes suffer from the light. They eat honey; death ensues should they eat apples of the mandrake unless they also devour ants. When fighting bulls bears seize their horns and inflict pain on the bulls' tender noses. The young are born head first, making this part weak and the arms and loins strong, whence bears sometimes stand erect. If wounded they cure themselves by touching the herb phlome or mullein. Numidian bears exceed others in fierceness and in length of shaggy hair.

Almost all of H's account is found in Pliny (viii.36.54) and earlier in Aristotle, who describes the new-born bear as "smooth and blind, and its legs and most of its organs are as yet inarticulate" (vi 579a 23), and who also mentions the eating of honey, ants, and fighting with bulls (vi 594b 5 f.).

The customary picture of the bear is that of a recognizable animal licking one or more round balls. In Bodl. 764, f. 22v. features are already visible in the mass beneath the bear's tongue.

BEAVER.

castor, fiber; castor, bievre.

The description of the beaver in both its text and illustrations is one of the most unvarying of all those in the *Physiologus*. According to B (17) and Y (36), the beaver is a very gentle animal in whose testicles is a medicine valuable for various maladies. When hunted, the beaver bites off its testicles and throws them at the hunter, who picks them up and ceases the chase. Should another hunter pursue the beaver, it shows its incompleteness and is left unharmed. The identical tale appears in all French bestiaries.

B-Is and H (ii. 9) change the report in no way, but add Isidore's etymology (xii.2.21) that *castores*, "beavers", are so called from *castrando*, "castrating". There are also beavers called Pontic dogs. The beaver's self-emasculation is recounted, among many other ancients, by Pliny (viii.30.47).

It is stated in the allegory that those who wish to live chastely should sever themselves from sin and throw it in the face of the devil, who will depart when he sees that nothing belongs to him.

Little variety appears in the illustrations of the beaver, which is portrayed as a dog-like animal oftentimes in extremely contorted positions. Usually there is a hunter in the scene (Pl. II, fig. 5). It is rare, but not unknown, to find the beaver looking like the amphibious animal it really is. Except for its yellow body and blue tail the beaver in B.N., fr. 14970, f. 13v. (GC) is realistically drawn.

BEES.

apes.

H (iii.38), like Isidore (xii.8.1), begins with an explanation of the name *apes*: bees are so called either because they hold to one another *pedibus* "with their feet" or because they are born without feet, only acquiring them later. They excel in making honey, dwell in the abode assigned to them, and arrange their home with great skill. They have an army and kings, wage war, flee smoke, and are excited by noise. It has been proved that bees are born from the carcasses of oxen. In order to create them the flesh of slain calves is beaten so that from the decayed gore worms are formed that afterwards become bees.

The remainder of H's long passage will be briefly summarized. Bees are the only animal to have everything in common - home, work, food, and even offspring. They elect their own king by choosing the most noble in body. The king does not use his sting vengefully. The laws of the bees are based on custom, and lawbreakers punish themselves, dying from their own sting. After speaking of the bees' ingenuity in the construction of their hexagonal cells, H enumerates their division of duties among which are watching over the food supply, examing the rain-clouds, forming wax from flowers, and collecting dew. Bees have a poison which they spread in their honey if they are irritated.[33]

In Pliny many chapters are devoted to bees and the production of honey (xi.5.4-20, 23). Although he says that the generation of bees has puzzled many, he does not offer the theory of spontaneous generation from carrion as a solution (xi.16.16), nor does Aristotle (v 553a 17 ff.), who quotes some as affirming that bees fetch their young from various flowers. Many details in H's account are also found in Ambrose's *Hexaemeron* (v.68.21-70).

Scenes of bee-husbandry are numerous, such as that in Morgan 81, f. 58 where bees fly into the hive from a cloth held by their keeper. In some examples they sting a nearby man as in Brussels, Bibl. Roy. 8340, f. 200v.

BLACKBIRD.

merula; *merle*.

The *Aviarium* (i.43), the only Latin account of this bird, cites Isidore (xii.7.69) for the etymology of the bird's name. Formerly the blackbird was called *medula* because it *modulet,* "sings", but others say that it was so named because it flies alone, as if *mera volans.*[34]

---

[33] Bees do not appear in French bestiaries until Richard de Fournival's *Bestiaire d'Amour* (37,10), where it is stated that although bees have no hearing, the swarm can be led by a fife or by singing, not because the sound can be heard but because the bee's nature is so orderly.

[34] *Merula* early gave rise to varied etymologies. Festus in the *De significatione verborum* defines *merum* as: "Antiqui dicebant solum; unde et

It is universally black except in Achaia.[35] The sweetness of the blackbird's voice moves the mind to a feeling of delight.

In PB (IV,81) the blackbird is described as an agreeable bird that sings marvelously high in April and May, and because of its song, people keep it in a cage.

Aristotle (ix 632b 16) notes the change in the voice of the blackbird which, he says, has a musical sound in the summer and in the winter a discordant chatter.

In most manuscripts of the *Aviarium* the blackbird is merely a black bird.[36]

BOA.

boa.

According to H (iii.45) the boa is an Italian snake of immense size. It follows herds of cattle and, fastening itself to the udders soaked with milk, sucks until it kills. From the depopulation of cattle, *bos*, it takes its name (Isidore, xii.4.28).

Pliny (viii.14.14) does not say that it kills the cow, but that it is nourished by the cow's milk.

Occasionally the boa will only appear as a coiled snake, but in Bodl. 764, f. 98 it is a winged serpent clinging to a cow's udder.

BOAR.

aper.

According to H (iii.17) the boar is a wild pig or hog so called because of its savageness, a letter being eliminated from the word

---

avis merula nomen accepit, quod solivaga est et solitaria pascitur", and Quintillian in his *De institutione oratoriae* (i.6.38) attributes to Varro (*De lingua latina*, v.76) the error of saying that *merula* is derived from "flying alone".

[35] Pliny says Arcadia (x.30.45).

[36] There is one unusual exception. The Hofer Bestiary, which is preceded by the *Aviarium*, has on f. 6 a miniature portraying a naked, tonsured monk walking through high grass, his clothes in a bundle nearby and a white bird atop a tree. This is one of the rare examples in a Latin manuscript of the moralization being illustrated. The story is from the life of Saint Benedict in the Dialogues of Saint Gregory (Migne, *Patr. Lat.*, LXVI, Col. 132). It is recounted that a blackbird flew before the face of Saint Benedict. The holy man was thereafter tempted by the flesh, and casting off his clothes,

asper, "wild", as Varro declares.[37] After presenting other etymologies unrelated to the boar, the notice ends by saying the name comes from the Greek ἀφρός, "foam", because it foams at the mouth.

BONASUS.

bonasus.

One of the few animals mentioned by H (iii.5) which is not found in Isidore but in Solinus (40.10) is the bonasus, an Asian-born beast. Its head is like a bull's and its horns curl in upon each other so that what they strike is not harmed. The forehead's lack of defense, however, is remedied by the belly, for when pursued, the bonasus expels dung in a swift flow for a great distance. The heat of the dung consumes whatever it touches.

The bonasus is described by Pliny (viii.15.16) who exaggerates a descripiton of the monapos by Aristotle (ix 630a 20 f.). Aristotle writes that the animal defends itself "by kicking and projecting its excrement to a distance of eight yards, ... and the excrement is so pungent that the hair of hunting dogs is burnt off by it."

The bonasus is sometimes given a horse-like body though with cloven hoofs, and the manner of depicting its involuted horns tested the cleverness of the artist. Occasionally a defenseless hunter stands behind the flow from this animal, but a wise one protects himself with a shield as in Camb., Fitzwilliam Musuem 254, f. 24.

BULL.

juvencus, taurus.

H (iii.18) says that the *juvencus*, "bullock", is so called because it begins to help (*juvare*) in cultivating the land, or because among the Gentiles it was sacrificed rather than the bull because the age of the victim was considered. The Indian bull is tawny colored, swift as a bird. Its hair turns contrariwise, its horns are movable, and its

---

he plunged into a thicket of nettles where by wounding his flesh he cured the wound of his mind.

[37] Varro (v.101) attributes its name to the fact that it frequents *aspera*, "rough" places, whereas Isidore (xii.1.27), substituting an *f* for a *p*, declares that the name is from its wildness (*a feritate*).

hard hide rejects spears. Such is their wildness that when captured, they breathe fiercely lest they be tamed. *Taurus* and *bos* are Greek names.

The Indian bull is mentioned in Pliny (viii.21.30), but the derivation of the word from *juvare* is given by Varro (v.96). In comparing H's text with Isidore's (xii.1.28) one sees that the former has omitted the phrase *jovi*, "to Jove", when speaking of the sacrifice of the bullock. This passage seems to be based on Servius (*Comm. in Verg. Aen.* III, 1. 21) from whom the sentence about the age of the victim is copied.

Differing from the usual picture of a lone bull is the miniature in B.M., Sloane 3544, f. 17, which shows a kneeling priest holding a tiny bullock in his hands beside an altar.

CALADRIUS.

caladrius, charadrius, calandrius; caladrius, caradrius, caladres, calandre

B (5) and Y (5) begin by referring to Deuteronomy (14:18), where it is written that the caladrius (heron in the A.V.) is not to be eaten. The first characteristic of this bird is that it is all white. Then a feature is described about which there is interesting divergence. B says: "cuius interior fimus curat caliginem oculorum". This is repeated in B-Is and H (ii.31), but in the *Aviarium* (i.48) the passage reads: "cuius femoris pars interior caliginem aufert ab oculis".[38] The bird is found in the king's house. It prognosticates the outcome of an illness, for if it refuses to gaze at a sick man, the man will die, but if it looks into the face of the patient, it draws the illness into itself and flies to the region of the sun where the sickness is consumed. DC and H vary the means of curing by saying that the bird puts its mouth upon the man's and with its breath extracts the illness. This, of course, is copied by G (863-878).

Allegorically the caladrius is Christ, who turned away from the Jews because of their disbelief and malice and turned instead to the Gentiles, whose infirmities and sins he bore to the cross.

---

[38] The variation in readings between *fimus* and *femur* is an understandable one, probably originally due to a copyist and then perpetuated.

The Latin model for PT (2143-2174) read *femur* since he writes:

E l'oisels at un os
Enz en la quisse, gros;
Se om la meüle at
Cui veüe faldrat,
E ses uiz en uindrat,
Senes repairerat.   (2167-2172).

The same is true for PB who speaks of the power of the bird's thigh.[39]

Of all bestiary subjects the identities of this bird and of the unicorn have probably aroused most speculation.[40] The origin of the legend of the caladrius has been investigated by Druce and it is his conclusions that will be summarized here.[41] In the list of unclean birds in Leviticus and Deuteronomy the Hebrew *anaphah* was rendered by the Septuagint as χαραδριός. This name is derived from χαράδρα, "mountain stream", which when swollen cuts its way through the mountain side forming a cleft, hence the cleft itself, and at last a bird dwelling in such a place. Aristotle speaks of the caladrius in this connection; elsewhere he places it among sea birds (viii 593b 15; ix 615a 1). There is no mention of its being white. Pliny (xxx.11.28) tells of curative powers like those of the caladrius belonging to the icterus, so called because of its peculiar color, icterus being the Greek word for jaundice. If the patient looks at this bird, he will be cured of jaundice, and the bird will die. Among the ancients, jaundice was known as *regius* or *arquatus*

---

[39] Arsenal 3516, f. 199v. of PB has a strange addition, which Cahier (II,129) reports as probably an interpolation since it is written above the line. The bird is described as having two straight horns like a goat's. In the illustration a later hand, according to Cahier but difficult to verify on the manuscript, has drawn two long horns from the bird's head. A small horned bird appears also in Montpellier H. 437.

[40] The caladrius has been identified as the plover, the lapwing, the crane, the woodcock, the parrot, and the heron. PT says it resembles a *mave*, "seagull". For other identifications see D'Arcy Wentworth Thompson, *A Glossary of Greek Birds* (Oxford University Press, 1936), pp. 311-14. In the Middle Ages there was some confusion between the caladrius and the *calandre*, the crested lark, because of a similarity of names. Littré gives the derivation of *calandre* as probably from *caliendrum*, a "high head-dress", linked with the crest of the bird. This information is stated in an article by George C. Druce, "The Caladrius and its Legend", *The Archaeological Journal*, LXIX (1912), p. 402, n.1.

[41] Druce, *op. cit.*, pp. 381-416.

morbus and *aurugo*. Druce concludes that the name "royal disease" led the *Physiologus* to say that the caladrius was found in the courts of kings.[42]

The depiction of the scene of the cure by the caladrius shows variations within a limited range. In the oldest Latin illustration, Bern 318, f. 8v. the white bird perches on the feet of a reclining woman with outstretched arms. Lines of yellow-white paint extend from its eyes to the infirm woman, thus assuring her of recovery. Brussels, Bibl. Roy. 10074, f. 142v. and 143 contains a drawing not found elsewhere of the capture of the bird with a net in the king's dwelling. Beside it a man holds the bird with averted head over the bed of the ill one. In a third scene this is repeated with the difference that the bird not only looks at the sick man, but in the same picture flies up to the sun. The figure of Christ is also present. Following the text, Morgan 832 of the *Dicta Chrysostomi* (Pl. III, fig. 3) shows the bird's beak against the patient's mouth before it flies to the sun. A crowned king looks at a web-footed bird standing on his coverlet in Morgan 890, f. 12. The French bestiary illustrations show only minor differences.

CAMEL.

camelus.

A lengthy explanation for the camel's name is given by H (iii.20) and is found in a slightly changed form in Isidore (xii.1.35). It may be derived from *cama* (κάμνω), since it is a beast of burden and chamae (χαμαί) is a Greek adverb meaning *humi*, "on the earth", for when burdened, camels recline and thus make themselves more humble (*humiliores*). Or it may be derived from *camur*, "curved", because of the hump on its back.

Arabian camels are numerous and have two humps, while Bactrian are the strongest and have only one hump, and their hoofs do not wear away. Some camels are suitable for bearing burdens and other for traveling. They grow wild with the desire to mate, and

---

[42] *Ibid.*, p. 410. A different explanation is offered by Wellmann, *op cit.*, pp.54-55. In the *Koiraniden* of Hermes Trismegistus, cited by Wellmann, the caladrius, whose healing power is not limited to a distinct illness, is described as ὄρνεον βασιλικόν; thus the "kingly bird" came to dwell in the residence of the king.

they hate horses. For three days they can endure thirst; then when given the opportunity to drink, they fill up for past and future needs. If muddy water, which they prefer, is lacking, they stir up clear water by treading. Their life span is one hundred years, but if they are taken to a foreign place they become ill because of the unaccustomed change of air. Females are used in war. Desire for mating is destroyed by castration for they are thought to be stronger if kept from engendering. Most of these observations are found in Pliny (viii.18.26) and Solinus (49.9).

It cannot be said that the camel's structure was too well understood by many bestiary illustrators. The merest trace of two humps appears in Morgan 81, f. 42, while the camel has horse-like legs in B.M., Harl. 3244, f. 47v. A man grasping a scourge sits sideways on a kneeling camel in Camb., Gonville and Caius 384, f. 178.

CAT.

cattus, musio, muscio, muriceps, murilegus.

Isidore only uses the words *musio* and *cattus* in his derivations (xii.2.38) on which H is based (iii.24). The *musio*, "cat", is so called because it is hostile to the mouse (*mus*). The people call it *cattus* from *captare*, "to seize". Others say it seizes things with its eyes —that is, sees— for so keen is its sight that it penetrates the night's shadows. The final sentence in H reads: "Catus enim acutus et callidus", and in Isidore: "Unde et a Graeco venit catus, id est, ingeniosus, ἀπὸ τοῦ καίεσθαι".[43] The latter reading is found in Servius' commentary on the *Aeneid* i.423.

A humorous view of the cat's relations with the mouse is often evident in the drawings of this subject. In Morgan 81, f. 46v. three cats —one green, one blue, and one tan— sit on their haunches; the tan one holds a mouse in its paws. In front of them a large blue cat stalks a fat mouse, which flees over the border of the illustration. A feline domestic interior is seen in Bodl. 764, f. 51.

---

[43] This Greek phrase, doubtless not understood by the scribe, is rendered in Bodl. 764, f. 51 as *apotoyragestai.*

CERASTES.

cerastes.

Again H's account (iii.42) is based wholly on Isidore (xii.4.18).[44] The cerastes is so called because horns extend from its head like a ram's, for κέρατα in Greek means "horns". Some have four pairs of little horns, which show when the rest of the body is covered by sand. These look like bait and attract animals, who are then killed by the cerastes. It is more flexible than other serpents, seeming to have no spine. These traits appear in Pliny (viii.23.35) and Solinus (27.28); Lucan (*Pharsalia* ix.716) speaks of the cerastes "which wanders about as its spine makes it turn". Aristotle (ii 500a 4) says that the word "horned" is applied metaphorically by the Egyptians to this animal, which actually only has protuberances on its head.

A vivid illustration in Bodl. 764, f. 102v. shows the head of the cerastes with its ram-like horns emerging from the sand while from its mouth hangs a dead bird.

CINNAMOLGUS.

cinnamulgus.

An Arabian bird according to H (iii.30), which repeats Isidore (xii.7.23), this bird is so called because it makes its nest from the fruits of the cinnamon tree. Men cannot reach the nest because of the height and fragility of the branches, so they attack it with lead balls to bring the cinnamon down. It is sold at a better price since merchants esteem this cinnamon more than any other.

A somewhat similar tale is recorded in Herodotus (iii.111) where old birds seize large pieces of meat and carry them to their nest, which the meat's weight causes to break and fall to the ground. The fable of the cinnamolgus as related by Aristotle (ix 616a 7), who says that men attach lead weights to their arrow tips to bring down the nest, is repeated by Pliny (x.33.50) and Solinus (33.15).

---

[44] The additional chapter on the Cerastes which concludes the version of the *Dicta Chrysostomi* in Munich, lat. 6908, f. 85v., is unrelated to the bestiary account. The drawing presents a prudent rider drawing up his leg as a horned snake bites his horse's foot.*

Illustrations commonly show this bird in its nest atop a tree while below a man holds a slingshot and a companion picks up the fallen cinnamon as in Bodl. 764, f. 71v. In B.M., Harl. 3244, f. 54 a figure shakes a club at the bird standing in a tree.

COCK.

gallus.

In his edition of Bern 318 Cahier omitted the chapter on the cock because the account was unique among the manuscripts that he had examined, and all lists of contents based on his edition have followed him in this omission. According to Helen Woodruff[45] this earliest Latin *Physiologus* account follows verbatim the notice of the *galli cantus* (which is the title of the Bern article) given by Ambrose in his *Hexaemeron* (v.24.88).

The *Aviarium* (i.36) does not follow Isidore (xii.7.50) in deriving gallus, "cock", from *castratio*, "castration", but quotes Gregory in his Moralia on Job[46] on the intelligence of the cock, which discerns the hours of the night and then crows. The *Aviarium* states that before crowing, the cock spreads its wings and strikes itself. This last trait, as well as the cock's ability to mark the passage of time, is recorded in Pliny (x.21.24). CUL's chapter differs from that of the *Aviarium* in repeating all of Isidore. This includes a statement that some say the cock's limbs are destroyed when mixed with liquid gold.

A single cock is usually portrayed in the act of crowing.

COOT.

fulica; fullica.

B (22) relates that this is a very intelligent bird which neither eats cadavers nor flies about but always stays in one place. B-Is (23) adds Isidore's explanation (xii.7.53) that the *fuliga* is so called because its flesh tastes like a hare's, for in Greek the name of this animal is λαγώς. The *Aviarium* (i.58) repeats B on the coot, then

---

[45] Woodruff, *op. cit.*, p. 253.
[46] Migne, *Patr. Lat.*, LXXVI, Col. 529.

copies the following from Isidore. The coot is a bird associated with ponds; its nest is in the middle of the water or on a stone surrounded by water. When it foresees a storm it flees to sport in its nest.

The allegory states that by the habits of the coot is represented the man who lives according to the will of God — not flitting around like a heretic or someone smitten with worldly desires and carnel lust, but remaining always within the catholic and apostolic church. In PT's slight variation the bird is the holy man who rejects flesh for self mortification while the nest represents the dwelling place of the holy man or hermit.

PT (2749-2764) follows the *Aviarium* closely except for saying that the bird plunges into the water when the weather is stormy. GC (1965-1986), where the bird is unnamed and whose illustration in B.N., fr. 14969, f. 37, showing birds with fish in their beaks, carries the legend "le oysel ke mangue le bon peison", adds that its flesh is like that of a heather hare. The non-*Physiologus* details of its appearance in PB (III, 208) remove the bird from the realm of precise ornithology: it is not big but is spotted with various colors; its head, beak, and feet are like an eagle's; its body is that of a peacock, and its neck and tail like a falcon's.

The drawings of the coot are usually an uninteresting portrayal of a bird on water or by a nest.

C RA N E.

grus; *grue*.

Cranes follow one another in lettered order.[47] They fly high to see more easily the lands they seek. The leader corrects his band with a shrill voice, and when he becomes hoarse, another succeeds him. At night cranes take turns keeping watch, the watchman hold-

---

[47] The *Aviarium* (i.39) does not cite here, as does Isidore (xii.7.14), a line from Lucan's *Pharsalia* (v.716) which would clarify what is meant by "ordine litterato": "Et turbata perit dispersis littera pinnis". In its context this is translated: "...at the beginning of their flight they [the cranes] describe various chance-taught figures; but later when a loftier wind beats on their outspread wings, they combine at random and form disordered packs, until the letter is broken and disappears as the birds are scattered". (Loeb translation). A note to this passage in the Loeb Classical Library Edition (p. 292) states: "Palamedes was said to have invented the alphabet by copying the figures formed by flocks of cranes in the sky".

ing pebbles in his raised foot and thus being prevented from sleeping. Their clamor is a warning. With age they become black. To this description as found in the *Aviarium* (i.39) CUL adds others. The notice here begins by repeating Isidore that the name *grus* comes from the sound of the bird's voice. To act as ballast in a strong wind cranes swallow sand and carry small stones. When one bird becomes tired, it is held up by the others until it recovers.

Most details in these accounts can be found in Pliny (x.23.30; x.29.42), Solinus (10.12-16), and Aristotle (ix 614b 18), who omits reference to the role of the stone but speaks of the leader's keeping watch. This is the only trait carried over to PB's bestiary (II, 142), where it is said that the pebble prevents the guard from standing firmly and thus going to sleep.

The characteristic pose of cranes as they are portrayed in the illustrations for both Latin and French bestiaries is that of the sleeping birds with their guardian holding a stone in his upraised foot (Pl. III, Fig. 2).

CROCODILE.

crocodilus, cocodrillus; cocodrille, cocatris.

Bestiaries of the Second Family contain two rather similar notices on the Crocodile, one between the chapters on the Leucrota and the Manticora and the other in the section on Fish. H (ii.8) repeats Isidore's etymology (xii.6.19), which states that the crocodile gets its name from *crocus*, its "saffron color". It is a four-footed animal, born in the Nile, living on land by day and in the water at night. It is generally twenty cubits long, and is armed with cruel teeth and claws. The hardness of its skin is such that its back is not hurt when struck by stones. After eating a man it always weeps. It is the only animal to move the upper part of its mouth and hold the lower part immobile. A person is made beautiful by smearing the dung of this animal on the face until sweat washes it away. H (iii.55) adds that the *serra* or sawfish can cut the crocodile's stomach.

Much of this account is in Pliny (viii.25.37), although before him Herodotus (ii.68) and Aristotle (i 492b 24) had mentioned the peculiar movement of the crocodile's jaw. The strange effect that this statement had on some of the illustrations of the crocodile will be later evident. The beneficial results of applying *crocodilea*, a substance found in the crocodile's intestines, is described in Pliny (xxviii.8.28). The weeping crocodile is spoken of by Asterius, bishop of Amasenus about 400 A.D., in a sermon entitled *In principium jejuniorum*.[48] He asks his listeners if they are imitating the crocodile, who weeps when he captures a man not because he is penitent but because the head is lacking in flesh and not suitable for food.

The French versions in their chapter on the Hydrus both repeat tradition and show fresh imagination in describing this creature unseen by the authors. PT (703-718) cites and translates much of Isidore. GC (1651-1684) believes that the crocodile somewhat resembles an ox and it appears thus in the illustrations. As for the ointment which removes wrinkles from the face, he says that several people still use it,

> Mes puisque la suor sorvent,
> Sachez, que nul preu ne lor tent.   (1683-1684)

The only consistency in the portrayal of the crocodile is that it usually has four legs; beyond that point the artist's fantasy is at play. The first surprise apparent in the early twelfth century Bodl., Laud Misc. 247, f. 152v. is that the head of the animal swallowing the hydrus is upside down.[49] The scene usually represented in the first chapter is that of the crocodile swallowing a man (Pl. III, Fig. 1). Other interpretations of this exotic beast deserve

---

[48] Migne, *Patr. Gr.,* XL, Col. 388.

[49] Druce, who made a study of illustrations of the crocodile and remarked on this feature in French manuscripts, said: "This feature does not, as far as I know at present, occur in any English bestiary". See George C. Druce, "The Symbolism of the Crocodile in the Middle Ages", *Archaeological Journal*, LXVI (1909), 318. His explanation for this peculiarity is that the artist might have been following an expression used by Pliny (viii.25.38) when describing how crocodiles are hunted by the race of men known as the. Tentyritae. The words applied to the crocodile are "hiantibus resupino capite ad morsum" ("faucibus" being understood), which correctly rendered would be "with its head turned back and jaws agape for biting". According to Druce the word *resupino* is also used in the sense of "reversed", and it perhaps was in this sense that the artist interpreted it (Pl. V, Fig. 2a). This seems a likely explanation also for the appearance in some manuscripts (among others Camb., Corpus Christi Coll. 53, [Pl. V, Fig. 2b]) of the crocodile lying on its back while a winged hydrus disappears down its throat.

mention here as examples of a certain type of mediaeval inventiveness. In B.M., Sloane 3544, f. 11 a finely dressed lady apparently is collecting *crocodilea* from a lionlike animal. The crocodile is an extraordinarily long donkey lying in water and surrounded by fish in Bodl., Douce 88, f. 12v., and a winged, cock-headed creature staring at its human victim in B.N., lat. 14429, f. 110v.[50] Finally in Camb., Gonville and Caius 372, f. 11 a small naked rider whips a running animal covered with scales. French illustrations also show wide variations.

CROW.

cornix.

A chapter on the crow is not found in all Latin bestiaries, and never in French bestiaries. Y (40), beginning with a verse from the Septuagint, "Hieremias propheta testatur quoniam: Sedisti sicut cornicola deserta" (Jer. 3:2), says that the crow belongs to one male only. If he should die, the female does not take another mate.

The monogamy of the crow is attested by the third century A.D. writer Athenaeus (*Deipnosophistae* x.394b), who cites Aristotle's comments on the ring-dove, which lives in solitary bereavement when its mate dies, as does the crow. Plutarch in his *Bruta ratione uti* (v.989A) says that Penelope's chastity was surpassed nine times by the female crow.

H (iii.35) omits entirely any reference to the crow's fidelity, and copies Isidore (xii.7.44). The crow is a long-lived bird called among the Latins by its Greek name *corone*. Augurs say that it reveals ambushes and foretells the future, but it is very wrong to believe that God entrusts his foresight to crows! By their voice they are said to predict rain (Virgil's *Georgics* i.388 is here cited).

A dark colored bird suffices for depicting the crow in most manuscripts.

---

[50] The influence of the spelling doubtless accounts for the small illustration in B.N., fr. nouv. acq. 13521, f. 25v. (PB) of the *coc codrille*, where a cock with a serpent's tail (the usual way of portraying the basilisk) approaches a dog.

DIAMOND and MAGNET.

adamas, lapis adamantinus, lapis magnis; adamas, diamant, aimant.

Because there was an eventual mingling of the properties and the names for the diamond and the magnet, the two are here treated together. In Y (24) the diamond is said not to fear iron or keep the smell of smoke. If found in a house, neither a demon nor other harm can arrive, and whoever possesses a diamond can overcome man and beast. In the second chapter on the Diamond (47) the stone is said to exist in the East, where it does not shine by day but is found by night. The diamond is so called because it overcomes everything.[51] The magnet (46) supports iron.

The allegory generally contains the Latin (or French) of the Septuagint version of Amos 7.7: "Vidi virum stantem super murum adamantinum et in manu eius lapidem adamantem in medio populi Israel" (B-Is 36). The diamond is Christ while the Eastern (*orientis*) mountain is God from whom all things come (*oriuntur*). Christ hid his descent from the celestial powers and came into the shadows of this world to illumine it. Neither death nor the devil nor any creature can prevail over Christ.

The diamond's being found only at night is thought by Wellmann to be the result of a transference of a similar comment by Epiphanius in the *De gemmis* concerning the carbuncle ($\check{\alpha}\nu\theta\rho\alpha\xi$).[52] The knowledge of the magnet's power of attraction is said to go back to Democritus.[53] The early Latin translation (Y) does not mention, as do later versions, the power of the he-goat's blood to dissolve the hard diamond, but such a statement is made by Jerome in his commentary on Amos,[54] where he quotes Xenocrates, a physician of the first century A.D. whose superstitious means of curing diseases were borrowed from Democritus.

In PT (2894-2922), under the name *adamas*, the stone is said to be split by goat's blood, to attract iron, and to glow by night on a mountain in the East. GC (3333-3362), containing the same mixture of traits, adds that with the fragments of diamonds one

---

[51] The relationship between the name of the stone and its function is clearer in the Greek (32): ἀδάμας δὲ λέγεται ὅτι πάντα δαμάζει.

[52] Wellmann, *op. cit.*, pp. 87-88.

[53] *Ibid.*, p. 88.

[54] Migne, *Patr. Lat.*, XXV, Col. 1073.

can cut gems, iron, and steel. The old belief in the stone's ability to avert evil is also recorded. These characteristics are all found in B-Is (36), H (ii.34), Isidore (xvi.13.2,3) and, among others in antiquity, Pliny (xxxvii.4.15). PB (IV,65) retains the simple account of the invincible diamond shining in the night.

An interesting variety of illustrations portrays the attributes and significance of the diamond (the magnet does not appear), from a shell-like object atop a mound (Bodl. 602, f. 32v.) and Christ upon a mountain (Bodl., Douce 167, f. 11), to the lively miniatures of GC, where miners wield pick-axes (B.M., Cott. Vesp. A. vii, f. 27v.) and a hooded man drops a diamond in a dish of goat's blood (Camb., Fitzwilliam Mus. J. 20, f. 67v.).

DOG.

canis; chien.

According to H (ii.17) there are several kinds of dogs: some for hunting wild beasts, some for hunting birds, some for guarding sheep, and others which watch over their master's house and property. They are said to be unable to live without men. A certain king, captured and held by his enemies, was rescued by a mass movement of his dogs. Friends of a slain man did not know whom to blame for the crime, but a dog disclosed the slayer in the midst of a crowd. A dog cures his wounds by licking, and a young dog bound to the patient is supposed to cure internal wounds. He eats little and returns to his vomit. When crossing a river and holding some food in his mouth, the dog sees his shadow; hoping to have the other piece of meat and opening his mouth, he loses what he holds. CUL begins with Isidore's etymology (xii.2.25), which says that *canis* comes from the Greek κύων; some, however, think it is from *canor*, the "sound" of barking.

The anecdotes recorded in H are amplified in CUL, where it is specified that it was a king of the Garamantes who was rescued by his dogs. This and similar manuscripts include references to Jason's dog, who died from hunger after his master's death, to King Lysimachus' dog, who threw himself on the king's funeral pyre, and to the loyalty of a Roman dog who followed his owner to prison and then tried to support his corpse in the Tiber. These accounts are all drawn from Pliny (viii.40.61), as is the reference to the *lycisci*, the offsprings of the wolf and dog. PB (IV,75) in general follows the less detailed chapter of H.

The Latin bestiaries often devote numerous scenes to the dog as in B.M., Harl. 3244, f. 44v. ff., where six illustrations appear. Among these are a dog standing by a drowsing shepherd, dogs chasing stags and rabbits, a dog howling over the pierced body of his master while another leaps at the neck of a standing man, and two dogs attacking and dragging the porter of a castle in the presence of a crowned king. A small illustration of PB (Montpellier H. 437, f. 247v.) shows a dog with meat in its mouth crossing water.

D O V E.

columba; colum, coulon.

Y (48) begins with the words of Saint John on the descent of the Spirit of God like a dove (John 1:32). The various colors of doves are then enumerated with special attention given to the red dove. B (31) says that the red dove rules over all and that it is this one which gathers the others into the dovecot.

The allegory perpetuated in the French bestiaries is that found in B, where it is recorded that God sent his Son in the form of a dove to gather mankind together in his holy church. Just as the dove has many colors, so were there divers manners of speaking through the laws and the prophets. Briefly, the colors and their meanings are: red, the predominant color because Christ redeemed man with his blood; black, obscure sermons; speckled, diversity of the twelve prophets: air colored, Elisha, who was snatched up through the air; ashen, Jonah preaching in hair shirt and ashes; gold, the three boys who refused to worship the golden image; white, John the Baptist and the whiteness of baptism; stephanite, Stephen, the first martyr.

The first eleven chapters of the *Aviarium* relate the dove and its parts to Christian dogma or institutions, as for instance Chapter V, which is entitled "A comparison of doves having red feet to the Church". Chapter XI, however, lists ten "properties" or attributes of the dove: its song is a mournful plaint; it lacks gall; it continually kisses; it flies in flocks; it does not live from plunder; it collects the better seeds; it does not feed on corpses; it nests in the holes of rocks; it sits on streams in order to see the shadow of the hawk, which it can then quickly avoid; and it has twin young. Most of these traits are found in CUL.

Aristotle (v 544b lf.) mentions various types of doves, and Athenaeus (*Deipnosophistae* ix.394) adds their colors. The prominence of the red dove can perhaps be traced to a tale told in this same passage of Athenaeus and more specifically in Aelian (iv.2). On Mount Eryx in Sicily is celebrated the festival of the Embarkation when Aphrodite is said to leave for Libya. At this time the doves, which are sacred to this goddess, disappear. After nine days a red dove flies forth from the sea and the others follow. The inhabitants then celebrate the festival of the Debarkation. The hawk's attack on the dove, but not the water episode, is elaborated in Aristotle (ix 620a 24).

PT (2389-2396) cites Isidore (xii.7.61) as saying that one dove attracts others to the dovecot, but this is not to be found in Isidore. In PB (III, 275) only the colors and their significance are given.

The portrayal of the dove separate from its appearance on the limbs of the Peridexion tree is a disappointing sight.

DRAGON.

draco ; *dragon.*

The dragon, the greatest of all serpents or animals on earth, is drawn from its hole into the air, which it stirs and causes to shine. Crested, it has a small mouth and an open tube through which both air enters and its tongue stretches out. Its strength is not in its teeth but in its tail, which harms by lashing. Its venom is harmless, but what is caught in its coils is killed. Not even the size of the elephant avails, for the dragon hides around the paths where the elephant usually walks, and binding the elephant's legs with its tail, it destroys by suffocating. Dragons live in Ethiopia and India.

This passage from H (ii.24) is very similar to that of Isidore (xii.4.4,5) beginning with the word *Draco maior,* which many of the later manuscripts copied. The crested dragon which kills the elephant is recorded by Pliny (viii.11.11-13) and Solinus (25.10-13), and it is the latter who describes its mouth and powerful tail (30.15). Pliny had noted that the dragon was destitute of venom (xxiv.4.20).

GC (2221-2238) is the only French bestiary to have a major chapter on the dragon, the others only speaking of it in connection with the enmity existing between it and the elephant. (Pl. III, Fig. 4a.)

The details of the illustrations of the dragon vary, as might be expected, but it always has at least two feet, wings, and a long tail.

It is often shown entwined around the elephant. The legend of the illustration in B.M., Harl. 3244 (Pl. III, Fig. 4b) reads: "De dracone ignivomo: Qui se in aerem iaculatur ut ipsum facit choruscare."

DROMEDARY.

dromedarius.

H's short notice (iii.21), taken from Isidore (xii.1.36), says that the dromedary is a type of camel but smaller and swifter, whence its name, for in Greek *dromos* (δρόμος) means "race" and "speed". It can cover a hundred or more Roman miles in a day. The dromedary chews its cud.

Usually drawn with two humps, a humpless dromedary in Camb., Trinity College R.14.9, f. 99 nonchalantly crosses its forelegs. The artist of Bodl. 764, f. 45v. took the subject of the dromedary as an occasion to portray what must be the Magi astride their mounts.

DUCK.

anas.

Isidore's etymology (xii.7.51) is used by H (iii.36), which repeats that the *anas*, "duck," takes its name from its *assiduitas*, "constancy," in swimming. Certain ducks are called *germanae* (Isidore's spelling) because they nourish others more. Pontic ducks feed on the poison of the sea and are not suitable for human sustenance. It is also thought that the duck gives the goose (*anser*) its name because the latter, although it lives on land, swims continuously.

A reference to the duck in Pliny (xxv.2.3) notes that the blood of the ducks of Pontus is used as an ingredient in antidotes since they derive their nourishment from poisons.

Illustrations do no more than portray a recognizable duck.

EAGLE.

aquila; aigle, aille.

Latin texts begin with a quotation from Psalm 102:5 (Vulgate): "Renovabitur ut aquilae iuventus tua." Since the following account has no parallel in ancient literature, it is believed that its formation was influenced by this citation.[55] When the eagle grows old, its eyes are covered with mist and its wings become heavy. It seeks a fountain, then flies above it into the region of the sun where its wings are burned and the mist consumed. Descending, the eagle plunges three times into the fountain and is wholly renewed. DC and TH also say that with age the bird's upper beak grows so that it hinders its eating. After the eagle strikes it against a stone, the beak is broken and the bird can eat again. Pliny (x.3.4) mentions the growth of the beak but offers no solution for its diminution, and Aristotle (ix 619a 16) states that the bird dies of starvation.

The *Aviarium* (i.56) and B-Is (8) repeat Isidore (xii.7.10) in deriving the eagle's name *ab acumine*, "from the sharpness," of its eyes. When the bird flies high above the sea, it can see fish in the water below and, descending like a whirlwind, captures its prey. This bird can gaze directly at the sun, and it holds up its young by the claws to see those which keep their vision motionless and are thus worthy of their kind. The offspring which turn their eyes away are cast out as degenerate. CUL here adds that the coot rescues the abandoned eagle and raises it. In Aristotle's account of the sea-eagle (ix 620a 2) a similar story of the testing of the young is found. Among others, Ambrose (*Hexaemeron* v.18.61) tells of the coot's nourishing the rejected eaglet.

PT (2013-2060) copies B-Is, and GC (657-1704) includes the sun-gazing test to discover the true offspring, if by chance the eggs in the nest had earlier been changed. The beak-sharpening is recorded first in French in G (828-848). In addition to the traditional account the strength of the eagle's eyes is such in PB's long version (II,164) that they hatch eggs merely by looking at them. After the eggs are laid, the eagle and its mate fly to separate trees and watch the nest for forty days and fast until the eggs open. This addition probably has a Biblical connotation.

The allegory as presented in B states that the man who is clad in old clothing and the eyes of whose heart are covered with mist should seek the spiritual fountain of God. Unless man be baptized and raise the eyes of his heart to the Lord who is the sun of justice, his youth will not be renewed. G adds that Christ is the rock on which the eagle sharpens its beak. Finally PT links the eagle with Christ, who came from on high to conquer men's souls as the eagle catches fish. The eagle looks at the sun just a Christ looks directly

---

[55] Lauchert, *op. cit.*, p. 10.

at his Father, and as the eagle lifts its young toward the sun, angels carry souls to God who receives only the worthy.

In the Latin versions more than one scene is usually illustrated as in Bodl. 764, f. 57v., where one eagle flies to the sun, another plunges into water, and a third holds a fish. Elsewhere (Pl. III, Fig. 5) the eagle tests its young and dips into the fountain.

ELEPHANT and MANDRAKE.

elephas; elefant, olifant.

Of all accounts in both Latin and French bestiaries that of the elephant is one of the longest, not only because of the numerous traits attached to this animal but because its connection with the mandrake is often included. For the earlier Latin version that of Y (20) will be summarized since it is more inclusive than B (33).

The elephant is reluctant to mate, but if offspring are desired he goes to the East near Paradise where the mandrake tree is found. The female eats first, followed by the male, and immediately she conceives. At the time of birth the female goes to a pool, which she enters up to her udders and there gives birth. B's explanation is clearer here: if birth were out of water, the elephant's enemy, the dragon, would devour the young.[56] The male stands guard to trample the dragon to death should it come. B states that the odor of the elephant's burning skin or bones will expel serpents from any place. Y contains yet another trait. Since the elephant has no knee joint, if he falls down it is impossible for him to rise. Thus he sleeps leaning against a tree. When a hunter wishes to capture him, he cuts the tree partially so that when the elephant leans against it, both fall. The fallen elephant cries out, but no large elephant can lift him. A small elephant must come and raise him with his little trunk.

The amount of material is so large that it necessitates two chapters in H (ii.25,26), but only one in B-Is (34). The first chapter repeats the earlier Latin; the second begins with incomplete etymologies from Isidore (xii.2.14-16). The elephant is so called in Greek (ἐλέφας) because of the great size of his body.[57] Among the Indians

---

[56] Bern 318 explains the birth in water as allowing the young elephant by swimming to reach the mammae of its mother.

[57] This is elucidated by Isidore, who says that the elephant has the form. of a mountain, for λόφος is "mountain" in Greek.

he is called *barrus*, whence the trumpeting is called *barritus*. His teeth are of ivory and his snout is called a *promuscis* because it collects food for his mouth. It is like a snake, fortified with a wall of ivory.[58] The ancient Romans thought elephants were Lucanian oxen: oxen because they saw no animal bigger, Lucanian because in Lucania Pyrrhus first opposed them in battle against the Romans. The Persians and Indians place wooden towers on them and fight with javelins as if from a wall.[59] They go in herds, salute men with whatever movements they can, and flee a mouse. The female bears her young for two years, only giving birth once. Their life span is three hundred years, and though they formerly were found in India and Africa, now they exist only in India.

At the end of this chapter H and B-Is cite Isidore for his information on the mandrake (xvii.9.30). It is named *mandragora* because it has *mala*, "apples", that are sweet smelling and the size of filberts. In Latin it is therefore called "earth apple", and the poets call it manshaped because the root has the form of a man. Its rind is mingled with wine and given to those whose bodies are to be operated on so that asleep they will feel no pain. There is a female mandrake with leaves similar to lettuce and fruit like plums; the male has leaves like a beet.

In the moralization the two elephants symbolize Adam and Eve who were in Paradise before Eve tasted the forbidden fruit and urged Adam to do likewise. They were then expelled into this world, which is like a pond because of its fluctuations and its countless pleasures and passions. Finally the incarnate Christ came and led mankind out of this lake of misery. The burning skin or bones of the elephant represent the works or commandments of God, which purify the heart and allow no evil to enter.

---

[58] According to George C. Druce, "The Elephant in Medieval Legend and Art", *Archaeological Journal*, LXXVI (1919), p.5, n.4, this is a mutilated passage from Lucretius ii.537:

> in genere anguimanos elephantos, India quorum
> milibus e multis vallo munitur eburno,
>
> in the breed of snake-handed elephants, which
> in their thousands provide an ivory palisade about
> India, (Loeb trans.).

[59] For a discussion of the elephant and its tower see William S. Heckscher, "Bernini's Elephant and Obelisk", *Art Bulletin*, XXIX (1947), 158-65.

Early accounts of the many traits attributed to the elephant are numerous. Ambrose, for instance, refers at length to the elephant in his *Hexaemeron*. Pliny (viii.5.5) speaks of the modesty of this animal which never couples except in secret. However, it is Wellmann's belief that the tale of eating the mandrake near Paradise is a direct borrowing of the story of the Fall of Man, which could have originated only with a Jewish writer.[60] The enmity of the dragon and the elephant is described in Pliny (vii.12.12), where its cause is attributed to the serpent's desire in the hot summer for the cold blood of the elephant. Although Aristotle (ii 498a) insists that the elephant does not sleep standing and that its back knees can bend, Diodorus Siculus (iii.27) says that the Ethiopians saw the tree of the sleeping elephant, and Strabo (xvi.4.10) states that its leg has a continuous, unbending bone. Most of H's and B-Is's statements are based on the first twelve chapters of Pliny's Book VIII, and some of the remarks on the mandrake are found in Book XXV(94.147-150) which Pliny largely took from Dioscorides.

The French accounts of the elephant are lengthy and contain many of the traits mentioned above, although PT is alone in including the episode of the sleeping elephant and the tree. In describing the elephant's trunk GC uses an interesting expression which must reflect his interpretation of its manner of functioning.

> L'olifant est mult corporu.
> Quant il vent en un pre herbu
> Hors de sa boche ist un boël,[61]
> Od quei il se pest el prael.   (3289-3292).

---

[60] Wellman, *op. cit.*, p. 41.

[61] *Boël* means "gut" or "intestine", from the Latin *botellus*, "little sausage". Evidently GC thought that the trunk could be ejected at will to aid the elephant in eating. Ambrose long ago gave a reason for this appendage (*Hexaemeron* vi.5.31): "Helephantus autem etiam prominentem promoscidem habet, quia, cum sit eminentior cunctis, inclinare se ad pascendum non potest".

In the Provençal poem by Richart de Berbezill, *Atressi com l'olifanz*, are the following lines:

> Atressi com l'olifanz
> Que, quan chai, no's pot levar,
> Tro que l'autre ab lor cridar,
> De lor voz lo levon sus,...

The word *voz* in the last line of the quotation has been thought to mean "voice", but it seems more probable that this also is a form of *botellus*, referring to the small elephant's lifting the fallen elder by means of its trunk.

In his section on the mandrake PT (1569-1612), after repeating Isidore's description of the plant, explains how to gather it. A dog that has been starved for three days is attached to the plant, bread is shown to him, and he is called from afar. The dog pulls out the root and dies from the shriek which it utters. Man must close his ears to its mortal shriek. From this root comes valuable medicine which cures all illnesses except death. In G's short account (381-404, 439-443) the mandrake is called *mandegloire*. No mention is made of the use of the dog in GC (3297-3332); it is therefore odd that the only French illustrations of the dog tied to a human-headed root are seen in manuscripts of GC (B.N., fr. 14969 and 14970). The bark of the mandrake boiled in water is good for sickness, and when any part of the body hurts, the crushed plant should be applied to that spot.

The account of obtaining the mandrake by means of a dog which is killed in the process is probably taken from a similar story told by Josephus (*Bel. Iud.* vii.180) about a plant called *baaras* which grows near the Red Sea, and of another told by Aelian (xiv.27) about a plant named *cynospastus* or *aglaophotis*.[62]

Illustrating the elephant's numerous attributes apparently intrigued the mediaeval artist, but apart from creating the general outline of this large animal, details are often presented in a curious way. That these animals had a trunk (of what form is often puzzling) and tusks (originating exactly where is also a problem) is often the limit of characterization. Later manuscripts picture the elephant carrying an elaborate castle (Pl. IV, Fig. 1b), while earlier ones are more varied in their scenes. Bern 318, f. 19 and 19v. shows an elephant wrapping its trunk around a tree representing a mandrake, at the base of which flows a small stream, the River of Paradise. On the ground lies a coiled serpent. The little elephant lifting its fallen elder is also seen. The early twelfth century Bodl., Laud Misc 247 (Pl. IV, Fig. 1a) depicts the elephants and the mandrakes as well as the birth in the water. Although the mandrake appears occasionally in the earlier Latin manuscripts either as a human form or as a human-headed plant, the picturesque portrayal of its gathering appears only

---

[62] A convenient list of the properties of the mandrake among the ancients is found in Charles B. Randolph, "The Mandragora of the Ancients in Folk-Lore and Medicine", *Proceedings of the American Academy of Arts and Sciences*, XL (1905), 487-537. For a more general history see C. J. S. Thompson, *The Mystic Mandrake* (London, 1934).

in manuscripts of GC (Pl. IV, Fig. 4) and in the drawings of Queen Mary's Psalter.

FIRE STONES.

lapides piroboli, lapides igniferi, terroboli, cheroboli;[63] *turrobolen, deus perres.*

According to Y (3) and B (3) fire stones, which are found in the East, are male or female. When far apart, there is no fire; but if they approach one another, a fire is ignited that burns everything. French accounts are similar.

The moral of this is that man should flee woman lest the good that Christ has placed in him be consumed. PT specifies that because love burns when men and women are close, monks and nuns are kept apart. He also states that Samson and Joseph were tempted by women with divers results.

Although sexual differences between stones were not unknown to ancient writers,[64] the above description of the fire stones is not recorded other than in the *Physiologus*.

Illustrations of fire stones usually show the bust of a man and a woman emerging from a mound surrounded by flames, or a couple holding stones.

FOX.

vulpis; vurpil, gupil.

The fox is a crafty and deceitful animal for when it is hungry and finds nothing to eat, it rolls in red earth to give the appearance of being covered with blood. It then feigns death by lying on the ground and holding its breath. Birds, seeing the fox's swollen and stained appearance and its tongue hanging out, believe it dead and sit upon it. The fox immediately snatches and devours them (B 15; Y 18).

H (ii.5) and B-Is (15), following Isidore (xii.2.29), say that *vulpis*, "fox", comes from *volupes*, "twisty-foot", because the fox never

---

[63] H (ii.19) says that the Greeks call these stones "chirobolos, id est manipulos".

[64] Cf. Pliny (xxxvi.21.39).

runs in a straight line but in circles. The capture of the birds is then recounted.

The fox in the allegory is said to resemble the devil, who pretends to be dead for those who live by the flesh until he has them in his jaws; the devil is in truth dead for those who are perfect in faith.

References to the cunning of the fox seem to derive more from the Bible than from Greek or Roman authors.[65] There exists, however, a similar story of the fox's catching birds in Oppian's *Halieutica* (ii.107). In Gregory's *Moralia* mention is made to the winding course of the fox.[66]

The influence of the *Roman de Renart* can be seen at the beginning of GC:

> Assez avez oï fabler,
> Coment Renart soleit embler
> Des gelines Costeins de Noës. (1307-1309)

as well as in the title to PB (II,207), *Li golpis Reinart*.

Over the centuries the picture of the supine fox and the careless birds has scarcely changed (Pl. IV, Fig. 2). Less common is the scene in B.M., Harl. 3244, f. 43v. where a fox with a fowl in its mouth is pursued by a club-wielding peasant.

FROG.

rana.

Y's chapter (44) is to my knowledge found nowhere else in this form in Latin bestiaries. Land-frogs are not troubled in the summer by heat, but they die if rain catches them. If water-frogs see the sun's rays and are heated, they dip into a fountain.

GOAT.

dorcas, caper, caprea; chevre, buc.

The goat likes high mountains and pastures in valleys. It sees long distances and can tell whether men are hunters or travelers (B 20, or as Y 21 says, whether the man is approaching with harm-

---

[65] Sbordone, *op. cit.*, p. 58.
[66] Migne, *Patr. Lat.*, LXXVI, Col. 96.

ful or friendly intentions). H (ii.13) and B-Is say that the *caper*,
"goat", is so called because it *captet*, "seeks", wild places, though
some say it is from the *crepitus*, "cracking", of its shin-bones. The
Greeks call the wild goat *dorcas* because it sees keenly.[67] These
derivations are from Isidore (xii.1.15-16).

Although GC moralizes at length about good works, both his
allegory and that of PT stem from parts of B, which says that like
the goat, Christ loves high mountains; that is, prophets, angels, and
patriarchs. As the goat pastures in the valley, so does Christ in the
church, where good works and alms are his food. The keen sighted-
ness of the goat signifies God's omniscience and his perception of
the devil's deceits.

The French bestiaries follow the Latin closely, but PT (581-592),
instead of stating the purpose of the man whom the goat sees, em-
ploys a circumlocution:

> [the goat] Tres bien set purpenser
> Se il deit luinz aler.   (589-590)

Two scenes depict the principal traits of the goat: either the
pasturing, where the goat often stands against a low tree and eats
from its leaves (B.M., Harl. 3244, f. 42), or the ability of the goat
to distinguish a man's purpose from a distance as in Morgan 832,
f. 6v., where a tufted goat stands on rocks while at his side a hunter
blows his horn and holds a leashed dog.

HE-GOAT and KID.—*hircus, haedus.* H (iii.16) devotes a chapter to the lascivious he-goat, which looks askance from lust, whence
comes its name since according to Suetonius *hirquii* are the angles
of the eyes.[68] Its nature is so hot that the diamond which neither
fire nor iron can break is dissolved by goat's blood. *Haedi* or *hoedi*,
"kids", are called from *edendo*, "eating", for they are fat when little

---

[67] In his homily on the Song of Solomon 2:9, which is given as "Similis
est fratruelis meus capreae, aut hinnulo cervorum in montibus Bethel",
Origen explains the word *dorcas*: "...*caprea, id est dorcas*, acutissime videt.
...dicimus quia dorcas, hoc est caprea secundum eorum physiologiam qui
de naturis omnium animalium disputant, ex insita sibi vi nomen acceperit.
Ab eo enim quod acutius videat... dorcas appellata est." Migne, *Patr. Gr.*,
XIII, Col. 56.

[68] This statement is from one of the lost works of Suetonius and is
cited in Servius' commentary on Virgil's Third *Eclogue* (1. 8), and in Isidore.
See C. L. Roth, *C. Suetoni Tranquilli* (Leipzig, 1924), Fasc. II, p. 302.

and of an agreeable taste, though others say that the *f* of *foedus*, "foul", was changed to *h*.

Illustrations of the he-goat merely portray a bearded goat with long horns.

GOOSE.

anser.

The goose makes known the night watches by its repeated cackling. No other animal smells the odor of man as it does, and by its crying the ascent of the Gauls to the Capitol was detected. Of the two kinds of geese, the wild fly high and in order; the domestic live in villages, clamor, and maim themselves with their beak (*Aviarium* i.46). The first part of this notice is taken from Isidore (xii.7.52) whose description is not unlike that of Ambrose's *Hexaemeron* (v.13.44) in its mention of the role of the geese during the attack by the Gauls. The sentence on the goose's sense of smell is copied from Servius (*Comm. in Aeneid* viii.652), who attributes his information to Pliny although it is not now found there.

Usually a recognizable goose is presented, but Bodl. 764, f. 83v. depicts a scene in which a wrathful goose cackles before a fox holding a limp chicken in its jaws.

GRIFFIN.

gryphes; gripon.

H (iii.4) describes the griffin as a winged, four-footed animal born in the Hyperborean mountains, which has a lion's body with wings and a face like an eagle. It is hostile toward the horse. Live men are torn to pieces by the griffin or are carried to its nest.

Herodotus first mentions the griffin (iii.116) but does not describe it. Pliny (vii.1.2; x.49.70) speaks of it twice, once calling it a winged monster noted for digging gold from mines, and again in his chapter on fabulous birds noting that griffins come from Ethiopia and have long ears and a hooked beak. Isidore's account (xii.2. 17) of their birthplace in the Hyperborean regions and their enmity toward the horse is taken from Servius' commentary on Virgil's eighth *Eclogue* (1. 27).

In PB (II,226) the griffin, designated as a bird living in the Indian desert which it leaves only to find food, has strength so great that it can carry a live ox and fly with it to its young.

The griffin is shown with a biting horse in Bodl. 764, f. 11v. (Pl. IV, Fig. 3) but often it appears holding a pig, a small boar, a ram, or an ox in its fore claws. In the bestiary of the Westminster Chapter Library, MS. 22, f. 25v. it holds the head of a prostrate man. In spite of the PB text mentioning no leonine characteristics in connection with the griffin, the drawing in Arsenal 3516, f. 205v. shows a typical eagle-lion but with empty claws.

HAWK.

accipiter.

A résumé of Isidore's account (xii.7.55,56), beginning "Accipiter, avis animo plus armata, quam ungulis", will be given first because many later bestiaries have copied it rather than that of the *Aviarium.* The hawk is more armed with determination than with claws, having great courage in a small body. Its name comes from *accipiendo,* "taking", related to *capiendo,* "seizing", for it is a bird which snatches greedily from other birds. On that account they call it *accipiter,* that is, a "robber". The hawk is said to be harsh with its young for when it sees them able to fly, it gives them no food, but beats them with its wings and pushes them out of the nest. It thus forces them while young to seek prey lest by chance they become lazy adults.[69]

In the *Aviarium* (i.13-19), the first chapter devoted to the hawk contains Gregory's commentary on Job 39:26, "Doth the hawk fly by thy wisdom, and stretch her wings toward the south?"[70] Gregory says that wild hawks loosen their old feathers by heating their limbs in the warm wind, or if wind is lacking, they make the air warm by beating so that with pores open either old feathers will fall out or new ones will grow. The next chapter tells of the two kinds of hawks: the wild, which catch and eat domestic birds, and the tame, which catch wild birds and return them to their master. In the remaining chapters (16-19) the use and training of hawks is

---

[69] This latter information is found also in Ambrose's *Hexaemeron* (v.18.59).
[70] Migne, *Patr. Lat.,* LXXVI, Col. 623.

briefly described as a means of depicting the righteous life which man should lead.

The hawk is sometimes portrayed on its perch, but more interesting are the contemporary scenes of hunting such as that in Bodl. 764, f. 76v. where a woman stands with a hawk upright on her hand while a man arouses the bird by beating on what resembles a cymbal. In a stream swim two ducks.

HEDGEHOG.[71]

herinacius, hericius, echinus; heriçon.

B (13) states that the hedgehog is said to have the appearance of a suckling pig. It is wholly covered with quills. At harvest time it goes into the vineyard, climbs a vine, and shakes the grapes to the ground. After descending and rolling on the grapes so that they become affixed to the quills, it takes the fruit to its young.[72] Y (16) adds that it resembles a ball. These descriptions are substantially like Isidore's (xii.3.7), which B-Is (13) includes and which say that the hedgehog is protected against danger by its quills. The moralizations all agree in declaring that man should care for his vineyard and spiritual fruits lest the devil carry them off.

H contributes the information (ii.4) that this animal when cooked is suitable for medicine. When the hedgehog senses a man, it rolls into a ball and protects itself with its quills - and creaks like a cart! To this CUL adds that the hedgehog is also called *echinus* (the name of the sea-urchin). When it realizes that the north wind will blow, it closes the northern of its lair's two breathing holes.

Plutarch (*De soll. anim.* 16.971F) mentions the hedgehog's manner of obtaining grapes and its ability to detect the direction of the wind; however, the passage in CUL is identical with that in Ambrose's *Hexaemeron* (vi.4.20). Pliny states that apples are the fruit collected (viii.37.56), and Aelian (iii.10) designates figs.

There are almost no changes or additions in the French accounts except for the vacillation between the types of fruit. According

---

[71] The explanation for the inclusion of the Hedgehog in the *Physiologus* and its usual position as the chapter following that on the Siren and Onocentaur will be found in the notice on the Siren.

[72] In recent years the *Illustrated London News* has contained an article on the feasibility of the hedgehog's collecting fruit on its quills. See Maurice Burton, "The Hedgehog and the Apples", *ILN*, Aug. 16, 1952, p. 264.

to GC (1113-1150) the hedgehog does not fear being taken by another beast, for how could a beast devour him when he would be so badly pricked? PB (II,198) adds a quaint note at the conclusion: "Et tot dis quant il va cargiés a ses faons, si va chantant".

Illustrations of the hedgehog are usually limited to its climbing a tree or vine or rolling on fruit (Pl. V, Fig. 1). A less common scene shows barking dogs attacking a bristling hedgehog (B.M., Royal 2 B. vii, f. 98).

HERCINIA.

hercinia.

This bird is named from the Hercynian forest in Germany, where it is born. Its feathers so glow in the dark that they make a visible path because of their brilliance (H iii.31; Isidore xii.7.31). The hercinia is briefly mentioned by Pliny (x.47.67) who only says that its feather shine at night like fire. It is Solinus (20.3) who enlarges the account and makes the bird a beacon to travelers.

In Morgan 81, f. 51v. the body of the hercinia is entirely covered with thick gold leaf and gleams forth from the vellum page.

H E R O N.

herodius, fulica, ardea.

The early compilers of the *Physiologus* wavered as to the attributes of this bird, which partakes of traits assigned to both the coot and the heron with the result that proper identification is almost impossible.[73] The presence of the *herodius* is due to one of the Vulgate readings of Ps. 103:17 with which Y (27) begins: "Herodion domus dux est eorum (hoc est fulice, in psalmo CIII)."[74] This bird is wise above all others and does not seek numerous resting places, but where it dwells it feeds. It does not eat carrion.

---

[73] D'Arcy Thompson, *op. cit.*, p. 102, equates the Greek ἐρωδιός for the most part with the heron, but says that it is a difficult word of varying and uncertain meaning.

[74] Jerome (Migne, *Patr. Lat.*, XXVIII) reads: "Ibi aves nidificabunt milvo abies domus eius", and the A.V. (104:17): "as for the stork, the fir trees are her home".

In B (22) this description is assigned to the *fulica*. Nowhere else in antiquity is the *herodius* so described.[75]

The *ardea*, "heron", is described in the *Aviarium* (i.47) as deriving its name from *ardua*, "steep places", because of its lofty flight (Isidore xii.7.21). It fears rain storms and flies above the clouds to avoid a tempest. When the heron flies off, it signifies a storm. It seeks its food in the water, but nests in high trees. The heron defends its young in the nest with its beak lest they be carried off by other birds. Some birds are white and others are ash colored.

A chapter on the heron does not appear in the section on Birds in all of the Second and Third Family bestiaries. Among the Latin writers who have mentioned the heron is Virgil, who in the first *Georgic* (l. 363) speaks of the bird fleeing before the onset of a storm, and Lucan (*Pharsalia* v.555), who notes its lofty flight.

In Bodl. 764, f. 64v. the heron stands on the surface of the water, an eel in its beak.

HOOPOE.

hupupa, epopus; huppe.

B (10) begins with two Biblical quotations: "Honora patrem tuum et matrem tuam" (Ex. 20:12); "Qui maledixerit patri et matri, morte moriatur" (Ex. 21:17). When the hoopoes' offspring see their parents grow old with feeble wings and eyes covered with mist, they pull out the old feathers, care for their eyes, and warm them under their wings. Once the parents have recovered they are grateful to their children, who tell them that it is in repayment for the kindness done to them in their youth. The brief moral is that children should honor and care for their parents.

After this model family scene a jarring note is added by the *Aviarium* (i.52) and B-Is (10), which have copied Isidore (xii.7.66). Isidore's passage is apparently derived from Jerome's commentary on Zechariah 5:9,[76] where the hoopoe is characterized as a very filthy crested bird which collects human dung and feeds on evil-smelling excrement, dwelling in it and around tombs.

Two ancient beliefs were fused in the account of the hoopoe. Although Aelian (x.16) says that the bird was venerated among

---

[75] PW, *op. cit.,* p. 1096.
[76] Migne, *Patr. Lat.,* XXV, Cols. 819-820.

the Egyptians for its love of its parents, this trait is usually associated among the writers of antiquity with the stork. Aristotle himself says: "It is a common story of the stork that the old birds are fed by their grateful progeny" (ix 615b 23). In Aristotle mention is also made of the hoopoe's constructing its nest out of human excrement (ix 616b 1).[77]

While stressing the care of its parents, PT (2575-2604) omits all reference to the bird's dirtiness, but does add, from Isidore, that if the blood of a hoopoe is rubbed on a sleeping man, devils will come and try to strangle him. In an attempt to reconcile the two contrasting natures of the hoopoe, GC writes:

> La hupe est un oisel vilein:
> Son ni n'est pas corteis ne sein,
> Ainz est fet de tai e d'ordure.
> Mes mult sont de bone nature
> Li oiselet, qui de li issent: (821-825)

G (989-1002) explains that the young lick the veils that have formed over their aged parents' eyes. The illustrations of the hoopoe are very similar; the young are shown pulling out the old feathers and feeding their parents.

HORSE.

equus, caballus.

Unlike other extant manuscripts of the *Physiologus* of an early date, Bern 318 contains a chapter on the Horse, which was omitted by Cahier in his edition of this work. Miss Woodruff[78] states that the text is like Isidore's (xii.1.42-48), in which certain phrases are taken from Ambrose's *Hexaemeron*.[79]

---

[77] In the footnote to this passage D'Arcy Thompson remarks that the nest is not made from human excrement, "but has a very offensive odour arising from accumulated excrement, as well as from a peculiar secretion of the bird. ... from this cause, and from the bird's seeking its food amidst dung ('avis obceno pastu', Plin. x.29.44), the hoopoe receives various characteristic epithets in countries where it abounds, e. g., Coq puant, Kothahn, Mistvogel, Stinkvogel, etc."

[78] Woodruff, *op. cit.*, p. 253.

[79] Only one similarity has been located by me (*Hexaemeron* v.9.25).

In all of the expanded Latin bestiaries the chapter on the horse is one of the longest, and because many of the details treat factual points about the horse, only an incomplete résumé will be given here. According to Isidore the name *caballus*, "horse", is derived from *cavando*, "making hollow", from the horse's hoofprint, though others call the horse *sonipes* because it makes sound with its feet. Horses are high-spirited, they scent war, they are stirred by the sound of the trumpet for battle and are aroused by an excited voice in racing. Grieved when they are defeated, they exalt in victory. Certain of them recognize the enemy in battle and attack by biting. Some know their masters and become disobedient if these are changed; others will carry none but their masters and when they are slain or are dying, the horses shed tears. Men who are to fight infer the outcome of the battle from the horse's sorrow or joy. In the centaur is mingled the nature of the horse and man. Isidore lists the regions where horses live over fifty years, and states that in Spain, Numidia, and France the span is thought to be shorter. He concludes by defining the four qualities to be sought in a horse: form, beauty, merit, and color.

The long chapter in H (iii.23) is preceded by an additional etymology: *equus* is derived from *aequus*, "equal", because horses are harnessed in two's or four's of equal size and they are coupled for strength and for racing. Certain horses associated with famous men are mentioned: Bucephalus and Alexander the Great; the horse of Caius Caesar, which allowed none but its master to mount; a slain Scythian king's horse, which killed the victor by biting and trampling; King Nicomedes' horse, which fasted to death after the king died; and finally a slain leader's horse, which killed itself and the victorious Antiochus when he tried to mount. The horse's lust can be destroyed by cutting its mane. On the forehead of the newborn colt is found a poisonous substance called *hippomanes* which is used in love potions. If it were immediately removed, the mother would not suckle the colt. The healthier the horse the deeper it immerses its nostrils while drinking. H then follows with a description of the qualities and kinds of horses. Ultimately all of the accounts including Solinus (45.5-18) take much from Pliny (viii. 42.64-66).

The illustration consists invariably of a single well-drawn horse.

HYDRUS, HYDRA, and ICHNEUMON.

hydrus, niluus, echinemon, hydra; *ydrus, idre.*

Although Y (38) calls this animal *niluus*, the word appears nowhere else, and *hydrus* is the name by which it is always known. And because B-Is (19) includes Isidore's comments on both the hydrus and the hydra (xii.4.22,23), eventual confusion arose concerning these serpents which were already none too clearly conceived, with strange results ultimately appearing in some illustrations.

According to B (19) there is an animal in the Nile called a hydrus (Y says it has the appearance of a dog) which is an enemy of the crocodile. Upon seeing a crocodile sleeping with its mouth open, the hydrus rolls in mud in order to glide more easily down its jaws. It enters the crocodile's mouth, is swallowed, and having torn the crocodile's viscera so that it dies, the hydrus comes out. Allegorically the crocodile signifies death and the hell to which the incarnate Christ descended and led out those imprisoned there.

Of the ichneumon Y (39) says that it is an enemy of the dragon. When the dragon is seen, the ichneumon covers itself with mud, closes its nostrils with its tail, attacks the dragon and kills it.

Much has been written on the hydrus, an animal hard to identify, and the ichneumon with which it has sometimes been confused because it was also the crocodile's enemy.[80] Wellmann[81] found that the ichneumon has been called ὕλλος, "ichneumon"; ὕδρος, "water snake"; ἔνυδρις, "otter"[82] or "water snake". The Roman historian Ammianus (xxii.15.19), after speaking of the tiny bird, the trochilus, which tickles the crocodile's cheeks, writes of the *enhydrus, ichneumonis genus*, which enters the crocodile's mouth. Wellmann also points out that ὕλλος and ἔνυδρος were the names in Asia Minor for pharoah's rat or the mongoose. Aristotle (ix 612a 17) describes the ichneumon plastering itself with mud when attacking the asp, which is its usual opponent, but Strabo (xvii.1.39) includes its enmity with the crocodile.

---

[80] See Robin, *op. cit.*, pp. 181-88 and Druce, "The Symbolism of the Crocodile...", *op. cit.*, 320-24.

[81] Wellmann, *op. cit.*, p. 14.

[82] In Camb., Fitzwilliam Museum 254, f. 27v, in the chapter on the crocodile and preceded by a section on the *enydros*, is a description beginning: "Est bestia quedam que vulgo luter dicitur..." The value of the otter's pelt is mentioned and its picture is drawn.

H (ii.7) and B-Is copy Isidore, and besides telling of the hydrus'
killing the crocodile, explain that it is a water serpent deriving its
name from ὕδωρ, "water". Its bite causes a swelling called boa
sickness because it is cured by ox-dung, *fimus bovis*. The chapter
continues with an account of the many-headed dragon which dwelt
in the Lernaean swamps. When Hercules cut off one of its heads,
three came in its place. Isidore comments "Sed hoc fabulosum est"
because Hydra is known to be a place vomiting forth waters which
devastate the country-side, and when one passage is closed, others
burst forth, so Hercules dried this place with fire and closed the
passage. The French bestiaries show no deviation from B in their
description of the hydrus which destroys the crocodile.

It is rare to see a hydrus in any form other than with its tail
projecting from the crocodile's mouth and its head emerging from
the side of the same animal (Pl. V, Fig. 2 a and b).[83] One of the
strangest pictures of the hydrus is that in B.M., Harl. 3244, f. 62
where a horned and goateed serpent swims over the water while
a man with spear and shield sits in the prow of a nearby boat.[84]
Sometimes the hydra has an illustration to itself like the five-headed
serpent in Morgan 81, f. 16; occasionally, however, the hydrus-
hydra confusion appears as in Sion College L $\dfrac{40.2}{\text{L } 28}$ f. 88v., where
the animal coming forth from the bear-like crocodile's side has
three heads.

HYENA.

hyaena, hiena, ẏena; hyene, luvecerviere.

B (18) recounts that there is an animal which in Greek is called
hyaena and in Latin *belua*, "beast". According to the law it should
not be eaten for it is dirty.[85] Jeremiah (12:9) is then quoted as

---

[83] In the Icelandic Bestiary the hydrus is called a bird and so it is
pictured, with feathers in the crocodile's mouth and a bird's head appearing
from its side.

[84] Druce, "The Symbolism of the Crocodile...", *op. cit.*, p. 323, sees here
the influence of Pliny's various references to the hydrus as a water snake,
and believes that man and serpent are seeking their common enemy, the
crocodile.

[85] The hyena is not specifically mentioned, but cf. Lev. 11:27 and
Deut. 14:8.

saying, " 'Spelunca haenae' hereditas 'mea facta est'."[86] The hyena is said to have two natures; it is sometimes male and sometimes female (Y 37). To this B-Is adds from Isidore (xvi.15.25) that this animal has a stone in its eye called *hyena* which, if held under the tongue, will enable one to predict the future.

The allegories in general derive from B, which says that the sons of Israel are like the hyena because first they served God and then they adored idols. Omitting this, GC develops the idea of the hyena's resembling untrustworthy, two-faced people. PT does not contain the traditional allegory, but briefly states that the hyena signifies an avaricious and lustful man whereas man should be stable.

Many additions are found in H (ii.10) besides that of the two-gendered animal. It is reported to live in the tombs of the dead and to devour their bodies. Solinus (27.23-26) is quoted as saying that the hyena circles houses at night uttering certain words and is thought by the occupants to be a man. He who goes out is eaten. After mentioning the hyena stone, H again cites Solinus as maintaining that this stone is found in the stomach of the hyena's young.[87] CUL includes even more details about the hyena. Because its spine is rigid, it must move its whole body in order to turn. If a dog crosses the hyena's shadow, it loses its voice.[88]

Twice Aristotle refutes the notion that the hyena has two genders, and accounts for this error with the observation that the peculiar structure of the reproductive organs of the male and female could deceive a casual observer (*De gen. an.* iii 757a 6; *HA* vi 579b 15). That Aristotle's accuracy went unheeded is seen in the repetition of the same error a thousand years later. The added traits credited to Solinus are found in Pliny (viii.30.44), who in his next chapter (45) describes the *corocotta*.

---

[86] Jerome's version differs from the *Vetus Latina* and reads "Numquid avis discolor haereditas mea mihi?"

[87] Although Solinus writes, "in quorum pupulis lapis invenitur", H states, "Veruntamen Julius Solinus refert de hoc lapide quod in ventriculis pullorum hyaenae invenitur". Probably there has been an error here in confusing *pupula*, "pupil of the eye", and *pullus*, "young animal", since the lapidaries and other bestiaries all locate the stone in the eye.

[88] CUL and a few other manuscripts conclude their chapter on the Hyena with a short description of the Crocote or Crotote (actually the Leucrota), the offspring of the hyena and the lioness. It too can imitate human voices. Apparently the only illustration of this animal included with the hyena is in B.N., lat. 11207, f. 5, where its parents are shown mating.

PT (1177-1214) says that the hyena is called *luvecerviere* in French, and mentions its double nature and the stone. The same material is contained in GC (1575-1606) who states that he does not know the animal's name in French.[89] In PB (III,203) the hyena is described as having the body of a bear, though of a different color, and the neck of a fox.

With few exceptions the hyena is portrayed eating a corpse or a human limb in a tomb (Pl. V, Fig. 5). A notably different scene is found in some manuscripts of DC (Morgan 832, f. 4; Munich lat. 6908, f. 79v.), where two standing hyenas embrace one another.

I B E X.

ibex.

The short account of the wild goat was one of the earliest additions to the First Family of manuscripts as represented by B-Is. The ibex is said to have two horns, whose strength is so great that if it jumps from the top of a mountain to the bottom, its whole body is held safe and upright by its horns (H ii.15). Isidore (xii.1.16) gives as the derivation of the plural form *ibices, avices,* because like "birds" they dwell in high places. Pliny (viii.53.79) mentions the ibex and its immense horns, but not their sustaining power.

Artists seem to have enjoyed portraying the plunging ibex. In B.M., Harl. 3244, f. 40v., pursued closely by a panting dog, the goat has lowered its horns prior to leaping, but in B.N., lat. 3630, f. 77v. it is shown upheld in a totally vertical position. An odd variation exists in Camb., Corpus Christi Coll. 22, f. 165v. where the two horns appear as elongated tusks.

IBIS.

ibis, ibex,[90] *hibicis; ibis, ibex, ciguigne.*

According to the law the ibis is the dirtiest of all birds because it feeds on corpses and remains at all times on the water's edge seeking dead fish or a cadaver. It fears entering the water because

---

[89] In B.N., fr. 1444, f. 247 of GC the hyenas's picture is labeled *De yelve.* This name becomes clear when in appears in Bodl. 912, f. 6 of GC as *yeine* — a hasty glance could easily confuse the two words.

[90] The words *ibis* and *ibex* have been confused in both Latin and French manuscripts.

it neither knows how to swim nor makes the effort to learn; thus it cannot go in the deep water where clean fish exist (B 14, Y 17). This information is repeated in the *Aviarium* (i.57), which only adds that snakes flee the ibis. B-Is copies Isidore (xii.7.33), who tells how the ibis inhabiting the Nile purges itself with its beak and feeds on snake's eggs.

Except for G, which differs slightly, the long allegories say in general that the ibis represents the miserable sinner who remains with the fruit of the flesh rather than imitating the good Christian who, having been reborn by water and the holy spirit, proceeds into deep waters, where he can find the numerous fruits of the spirit such a charity, joy, and peace.

The unclean eating habits of this bird, sacred to the ancient Egyptians, are mentioned by Aelian (x.29). According to Pliny (x.28.40) the Egyptians invoke the ibis against incursions of serpents. In his chapter on medicinal remedies which have been borrowed from animals, he also reports that the ibis purges itself (viii.27.41).

In PT (2631-2742) the ibis, wrongly called *ibex*, is equated with the stork (Strabo had noted the striking resemblance between the two birds (xvii.2.4)):

> Ibex d'oisel est nuns
> Que ciguigne apeluns;
> D'Egypte vient del Nil,
> Mult par est best vil.   (2631-2634)

GC (1171-1194) confesses ignorance as to the bird's name in French.

Usually portrayed as a nondescript bird, the ibis is shown in many positions. One drawing combines the two principal traits of its self-purgation and its carrion diet, Bodl. 602 (Pl. V, Fig. 4). In Bodl. 764, f. 14v. a long-necked bird feeds its young in a nest on the ground (in some illustrations the food consists of snake's eggs) while under foot at the water's edge it holds a snake and a human head.

INDIAN STONE.

lapis indicus, lapis senditicho

Y (26), A and C, the only Latin versions to include this stone, say that if it is bound to a dropsical man, it absorbs his impurities, and when untied it weighs as much as the man. If set in the sun for three hours, the unclean water flows out and the stone is again pure.

Although the Indian stone was known in antiquity, being mentioned by Pliny (xxxvii.10.61) among others, this tale was not attached to it. Wellmann sees a parallel with the description of the water stone found in the head of a water snake as told in Hermes' *Koiraniden*,[91] but Sbordone thinks there exists an even closer relationship in this same work with the account of the land frog which carries in its head a stone capable of curing dropsy if fastened around the waist of the sufferer.[92]

The only illustration of this stone that has been seen is that of Bern 318, f. 21, where a seated man holds towards the sun a rectangular stone from which water drips.

JACKDAW.

graculus.

The *Aviarium* (i.45) attributes the beginning of its chapter on the Jackdaw to Rabanus Maurus, the ninth century author of the *De universo mundo*.[93] The *graculus*, "jackdaw", is thus called because of its *garrulitas*, "chattering", not, as some say, because it flies *gregatim*, "in flocks". The jackdaw lives in the forest and flies from tree to tree singing noisily. When caught it is shut in a cage to be taught to speak. If the bird escapes and was previously voluble, after leaving it shouts even more. Isidore (xii.7.45) contains the same statement as Rabanus.

The only representation seen other than that of an undistinguished bird was in Bodl. 602, f. 64v. where a small archer shoots an arrow into a large bird on a tree.

JACULUS.

jaculus.

H's chapter (iii.46), which is copied from Isidore (xii.4.29), states that the jaculus is a flying serpent about which Lucan wrote "Jaculi-

---

[91] Wellmann, *op. cit.*, p. 91.

[92] Sbordone, *op. cit.*, p. 141.

[93] Migne, *Patr. Lat.*, CXI, Col. 247. The chapter on the Jackdaw is not found in all Second Family manuscripts.

que volantes" (*Pharsalia* ix.720). It lies in ambush in a tree until some animal is nearby, then throws itself (*jactant se*) upon the animal, killing it, whence comes its name. Pliny (viii.23.35) and Aelian (vi.18) both mention this serpent.

In B.N., lat. 3630, f. 94v. a winged serpent darts from a tree to seize a naked man by the neck.

KINGFISHER.

halcyon.

The kingfisher is a sea bird, according to H (iii.29), which gives birth to its young on the shore, laying eggs in the sand around mid-winter when the sea surges. An unexpected calm then takes place. For seven days the kingfisher hatches its young, and for seven more days it nourishes them. Sailors watch for these fourteen days, during which no tempest is feared. This account is almost identical with that found in Ambrose's *Hexaemeron* (v.13.40) and differs in form from Isidore's (xii.7.25), which says that the name *alcyon* is like *ales oceanea*, "sea bird".

Aristotle (v 542b 4) speaks of the fourteen halcyon days at the winter solstice, and Pliny (x.32.47) of course includes this bird and its attractive story.

Consistent with this bird's close association with water, the artists have usually given it webbed feet as in B.M., Harl. 3244, f. 53 where the bird leans over its eggs. In Bodl. 764, f. 69v. the length of the bill of the bird standing on water and biting its tail is far from reality.

KITE.

milvus; escouffle.

The *Aviarium* (i.40) cites Isidore (xii.7.58) as its source for the statement that the *milvus*, "kite", takes its name from its being *mollis*, "supple", in strength and flight. It is exceedingly rapacious and feeds on carrion. Continuously flying around kitchens and meat markets, it quickly seizes any raw meat that is discarded. It is timid in large undertakings, bold in small. Not daring to snatch woodland birds, it lies in wait for domestic birds and young ones to catch and kill. PB (IV,82) is the only French bestiary to include the kite, whose dirty eating habits he dwells on at length. None of these descriptions follows Pliny (x.10.12) closely, although he speaks of the ravenous appetite of the bird.

The kite with its prey is the subject usually illustrated as in the *Aviarium* of Bodl. 602, f. 61, where the bird flies off with what resembles a young dog in its claws.

LAMB.

agnus.

The lamb is so called because ἀγνός means "chaste", although the Latins think that it is because the lamb *agnoscat*, "recognizes", its mother before other animals. Even if wandering in a large flock, the lamb immediately knows by their bleating the voices of its parents, just as its mother knows its offspring among many lambs (H iii.15).

The etymologies in this account are taken from Isidore (xii.1.12), the former being mentioned by Festus. Varro had said (v.99) that the lamb's name came from the fact it is *agnatus*, "born as an addition", to the flock of sheep.

Most drawings of domestic animals in the bestiaries show an acquaintance with the animal depicted, like the blue and green lamb in Morgan 81, f. 40.

LEUCROTA.

leucrota, leucrocuta.

This swift animal is born in India. It is the size of an ass, has the hind quaters of a stag, the chest and legs of a lion, a horse's head, cloven hoofs, a mouth split as far as the ears,[94] and instead of teeth, a continuous bone. With its voice it imitates the sounds of a speaking man (H iii.7).

The illustration in CUL, f. 14v. attempts to fulfill the description of this animal by giving it a toothless mouth which opens to its ears.

---

[94] The manuscript used for Book III of the *De bestiis* of Pseudo-Hugo (Migne, *Patr. Lat.*, CLXXVII, Col. 85) reads *ad nares*, but other manuscripts as well as Pliny (viii.21.30) and Solinus (52.34) read *ad aures*, "to the ears".

LION.

leo; *leun, lion.*

Since the lion was considered king of beasts, it is with its description that every Latin and Old French bestiary began until the unorthodox *Bestiaire d'Amour* of Richard de Fournival in the mid-thirteenth century. From the simple enumeration of three characteristics in the earliest Latin versions, the chapter became one of the longest and most complex. This increase is also seen in the number and subjects for the illustrations of the lion.

The traditional beginning for the first chapter of the early Latin *Physiologus* is that of B (1): "Et enim Iacob, benedicens filium suum Iudam, ait: Catulus leonis Iudas..." (Gen. 49:9). The lion is then said to have three natures: when walking in the mountains and smelling a hunter, it covers its tracks with its tail; it sleeps with open eyes; its cubs are born dead and on the third day it breathes into their faces and revives them. A slight variation in the manner of resuscitation is found in TH, where it is the roaring of the father which arouses the cubs.

Although PT has added numerous details to correspond to his lengthy description, basically the allegories of all Latin and French versions are the same: Christ everywhere covered the traces of his divinity; his body might sleep, but his divinity is ever watchful; his omnipotent Father revived him on the third day.

In ancient literature the erasing of the tracks by the lion's tail is not attested,[95] but this trait can be compared with Aelian (ix.30), who says that when the lion returns to its den it obliterates its path by running about. Among the classical writers who mention the open eyes of the sleeping lion is Plutarch (*Quaest. conviv.* iv.5.2). No writer preceding the *Physiologus* says that the cubs are born dead, but this idea, with its obvious later Christian connotation, could derive from Pliny (viii.16.17), who quotes Aristotle (vi 579b 8) as saying that the cubs when first born are shapeless and very small.

H (ii.1) and B-Is follow Isidore (xii.3.4) in saying that *leo* comes from the Greek, and they copy many traits from him and from Solinus (27.13-16). They state that there are three kinds of lions: the timid with a short body and curly mane; the more fierce with

---

[95] PW, *op. cit.,* p. 1078.

a long body and straight hair; the third kind is not described. Pliny (viii.16.18) speaks of only two species — those mentioned above. Their courage is apparent in their forehead and tail, their strength in their chest, and their firmness in their head. When surrounded by hunters they look at the ground, which frightens them less than the sight of spears. They fear the noise of wheels, more so fire, and above all the white cock.[96] The *Physici* is then quoted as saying that the lion has five natures. In addition to the usual three are added the traits that the lion is not easily angered unless first annoyed, that it spares the prostrate and permits captive men whom it meets to depart, and that it does not kill except out of great hunger. CUL adds other attributes not found in H and B-Is. A sick lion is cured by eating a monkey. It eats one day and drinks the other, but when it perceives that meat is not properly digested, its nails pull it from its throat (Pliny viii.17.18). From Ambrose's *Hexaemeron* (6.6.37) comes the statement that scorpions harm the lion if it wounds them and snake venom kills it. CUL continues with statements about the lion's mating practices and the belief that the lioness first has five offspring, then one less each year. This version concludes with a reference to the *leontophonus*, a beast whose ashes are spread on meat in order to catch lions (Isidore xii.2.34).

All of the Old French bestiaries keep the principal traits of the tail, eyes, and cubs, but some of them retain more and also contribute new details. PT (25-370) says that when the lion is hungry it treats animals angrily as it does the ass. When the lion wishes to capture prey, it makes a circle on the ground with its tail. Once inside, no animal dares pass beyond its limits. PT contains one trait which is as unclear at first sight as its illustration.

> Leüns quant est iriez
> Il se peint od ses piez,
> En tere se peindrat
> Quant il mariz serat;   (121-124)

The Latin rubric in both the Merton College and the Copenhagen manuscripts reads: "Hic leo pingitur et quomodo pingit se supra

---

[96] Lucretius ingeniously explains the fear of the lion before the cock. "... no doubt because there are certain seeds in the cock's body, which when they are sped into the eyes of the lion, dig holes in the pupils and cause stinging pain, so that they cannot endure against it for all their courage." (Loeb trans. *De rerum natura* iv.710-717).

pectus hominis." In the former the lion is shown standing over the body of a man lying in bed; in the latter it is standing on the chest of an animal lying upside down. The matter seems to have been clarified by Langlois who explains that *paindre*, "to strike", should be read instead of *peindre*, "to paint".[97]

In G the influence of a Biblical quotation (Song of Songs 5:2) is evident in the wording.

> Quant lion dorment et someillent
> Lor cuer dorment et lor oel voillent.
> Dex dist une mult grant mervoille:
> "Je dor" fait-il, "mais mes cuer voille".
>
> (109-112)

PB (II,110) has a full account of the lion, adding the strange statement that the cub is born dead through the mouth. He develops the point, which is continued in later bestiaries, that if a man should look at the lion while it is eating, because the man is made in the image of the Lord of Lords, it fears his face and glance. Since it is naturally bold, the lion is afraid of its fear and thus leaps upon the man. Had the man not gazed, he would not have been harmed.

Drawings of the several traits of the lion are more numerous than those of any other animal. Usually more than one picture is devoted to it, and in the finer Latin manuscripts there are as many as two entire folios portraying various aspects of the lion. That an actual lion had not been seen by the majority of artists is evident from its dog-like appearance, but the fullness of the mane and the length of the claws show that the artist knew the main physical characteristics of the animal.

The oldest Latin illustrated manuscript, Bern 318, f. 7 has as its first illustration one that seems to be unique in the *Physiologus* series. Jacob is shown blessing the lion of Judah, whereas in Brussels

---

[97] Langlois, *La vie en France au moyen âge* (1927 ed.), Vol. III, p. 15, n. 3, referring to Walberg's edition of PT writes: " 'Voilà, dit l'éditeur (p. 121 - Walberg), un trait bien curieux [faire son portrait avec ses pieds]. Il semble plus naturel de lire *paindre* (<pangere), c'est-à-dire *frapper*, au lieu de *peindre*; mais, comme les rubriques latines [de quelques mss.] indiquent que, pour leur auteur, il s'agissait bien de *peindre*, je n'y ai rien changé'.

Il fallait changer. Et l'erreur évidente des rubricateurs en cette circonstance est un argument qui s'ajoute à tous ceux que M. Mann a fait valoir (*Romanische Forschungen*, VI, 1891, p. 399) pour établir que les rubriques latines des manuscrits sont, non de l'auteur, mais d'un copiste."

10074, f. 140v. Jacob is blessing his son Judah. The Bern manuscript then has three separate pictures showing the lion dragging
a long tail, hunched up and sleeping with open eyes, and an open-
mouthed lion standing over a dead cub. These are the most commonly reproduced scenes. A more elaborate composition is seen in
Camb., Corpus Christi Coll. 53 (Pl. VI, Fig. 1b), where the lion
covers its tracks, revives its young, gazes at a scorpion, spares a
prostrate man, and eats a monkey. An unusual addition is seen in
the Third Family manuscript, Camb., Fitzwilliam Museum 254, f. 15,
where an episode from the story of Androcles (called *Andronicus*
in the text) and the lion is illustrated:[98] the lion is licking a naked
man tied to a stake. Although the text above the miniature in
Bodl. 764, f. 4v. tells of the *leontophonus*, the picture itself shows
how to capture a lioness. A sheep is bound to a log which projects
from a hole while down another hole, to reach its prey, the lioness
disappears.

The two illustrated manuscripts of PT have a record seven
drawings of the lion. The drawings of one are sometimes needed
to understand the subject of the other in spite of the Latin rubrics,
since these are often placed in the text at some distance from the
illustration. The scenes shown are the lion attacking an ass, driving
assorted animals into a circle, standing over a man or animal, cover-
ing its tracks with its tail, in the presence of a cock and a cart
(Pl. VI, Fig. 1a), standing by the dead cub, and Merton College 249,
f. 2v. contains a drawing of an animal before the bust of a man
holding a trilobed mace.

LIZARD.

saura, lacerta.

B (37)[99] calls this a "volatile animal", while DC (12) says that
it is an "aquatile animal". They agree that it is as bright as the
sun, whence the name *saura eliace* in Y (49). When it ages and

----

[98] Cf. Aulus Gellius, *Attic Nights*, v.14.

[99] In his edition of *Versio B* Professor Carmody is in error when he
publishes the *Lacerta* as the final chapter of this version. This reptile never
belongs to the group of First Family manuscripts known as the B version.
On the other hand, it is always included in the small, easily identifiable
family represented by Book II of Pseudo-Hugo, and in the *Dicta Chrysostomi*
manuscripts.

its sight no longers permits it to see the sun's light, it seeks a wall facing east. After passing through a hole and staring at the sun, its eyes are renewed (Isidore xii.4.37; H ii.28). The similarity of this mode of renewal with that of the serpent and its old skin is marked. There is no Old French version.

The lizard as portrayed by the mediaeval artist seldom has markedly reptilian features; it is usually a dog-like animal squeezing through a crevice. The example from Munich, lat. 6908 of DC (Pl. V, Fig. 3) is unlike any other representation seen.

LYNX.

lynx.

According to H (iii.3) and Isidore (xii.2.20) the lynx is so called because it is classed in the species *lupus*, "wolf", which it resembles except for being spotted on the back like the male panther (*pardus*). The urine of the lynx hardens into a precious stone which Pliny and Solinus call *lygurius* or *lycurius* (Isidore writes *lyncurius*). This animal covers its urine with sand lest the stone pass into human hands. Pliny is quoted as saying that the lynx has only one offspring.

The stone which is produced by the urine of the lynx and whose name means "lynx-water" is described by Pliny (viii.38.57), who says it is similar to a carbuncle and of a brilliant flame color.[100]

The lynx is usually shown standing over a round stone as in Bodl. 764, f. 11.

MAGPIE.

pica.

In H (iii.32) the magpie (*pica*) and the woodpecker (*picus*) are included in the same chapter, but they will be treated separately

---

[100] In another place Pliny asserts his disbelief in the existence of such a stone (xxxvii.3.13). Strangely no mention is made in the traditional bestiaries of the proverbial keen eyesight of the lynx (Pliny xxviii.8.32). Such a reference does appear in Brunetto Latini's chapter on wolves (*Li Livres dou Trésor* i.190). The metamorphosis of the lynx into a worm symbolizing the sense of sight is mentioned in the Appendix of the present study in the section entitled Elements and Senses.

here. Magpies are like poets because they express words distinctly like men.[101] Hanging from branches, magpies chatter annoyingly. If they are unable to speak, they can at least imitate the sounds of the human voice. It is fitting therefore that a certain poet (Martial xiv.73) should say:

> Pica loquax certa dominum te voce saluto.
> Si me non videas, esse negabis avem.[102]

The illustrations of the magpie are usually undistinguished.

MANTICORE.

manticora.

The manticore is born in India. It has a triple row of teeth which fit alternately, a man's face, bluish eyes, a lion's body the color of blood, and a tail like the sting of a scorpion. Its whistling voice resembles the melodies of pipes. It seeks human flesh, is active, and leaps so that neither large spaces nor broad obstacles delay it (H iii.8; Solinus 52.37).

Most early writers name Ctesias as their source for this fabulous animal.[103] Aristotle (ii 501a 24 f.), following Ctesias, speaks of the spines in the animal's tail which it shoots off arrow-wise, but Pliny omits this detail (viii.21.30).

Artists varied in their portrayal of the head of this beast, in one case giving it a woman's head (Camb., Gonville and Caius Coll. 384, f. 175v.), but more usual is its depiction as a heavily maned beast

---

[101] According to a note on this passage in Migne's edition of Isidore (*Patr. Lat.*, LXXXII, Col. 465), the phrase "picae, quasi poeticae" is reminiscent either of a verse of Persius (*Choliambus*, 1. 13) where he speaks derisively of "corvos poetas poetridas picas", or refers to the fable of the Pierides, the nine daughters of Pierus who were defeated by the Muses and turned into magpies (Ovid, *Met.* v.300-678).

[102] "A chattering pie, I with intelligible voice salute you, my master; did you not see me you will say I am no bird." (Loeb trans.)

[103] The account from Ctesias' *Indica* can be found in the *Myriobiblon* of Photius (Migne, *Patr. Gr.*, CIII, Col. 214), where the beast is called a μαρτιχόρα and is equated with the Greek ἀνθρωποφάγος, "man-eater". According to the *NED* the word manticore is derived from an Old Persian word *martijaqâra* also meaning "man-eater".

having a man's face topped by a Phrygian cap as in Bodl. 764 (Pl. VI, Fig. 2). How to depict a scorpion-like tail concerned none of the artists except the literal minded illustrator of B. M., Harl. 3244, f. 43v., where an oddly pointed tail is attached to a sharp toothed monster clawing a human body.

MOLE.

talpa; talpe.

The mole is a small black animal condemned to perpetual blindness for it lives in darkness without eyes. It is always digging in the earth, piling up soil, and eating roots (H iii.26; Isidore xii.3.5).

Although untrue, the belief in the mole's blindness has persisted from early times, both Aristotle (i 491b 27) and Pliny (xi.37.52) mentioning it. The remainder of the above description is taken almost verbatim from Jerome's commentary on Isaiah 2:20.[104]

In PB (III,274) the mole is twice cited: as a symbol of one of the four elements, earth, on which it is said to live; and in its own chapter, where the mole's ability to hear so well that it cannot be surprised is noted. According to PB it has eyes beneath the skin, it dirties the earth where it dwells, and harms grass.

The illustration of this animal in Latin bestiaries is usually a dorsal view of a flattened mole as in Morgan 81, f. 47.

MOUSE.

mus.

The mouse is a small animal whose Greek name ($\mu\bar{\upsilon}\varsigma$) was transferred into Latin. Some say that *mus* comes from its being born from *humus*, "soil". Its liver increases with the full moon—a belief found in Pliny (xi.37.76). Some make a division between the mouse and the shrew-mouse (*sorex*), which is smaller and pursued by cats (H iii.25; Isidore xii.3.1,2).

Mice are usually drawn with a round object, probably grain, in their mouth as they appear in Morgan 81, f. 47.

---

[104] Migne, *Patr. Lat.*, XXIV, Col. 55.

NIGHTINGALE.

luscinia; *lousegnol, rosignol.*

The *luscinia*, "nightingale", according to Isidore's etymology (xii.7.37) which is copied from Ambrose's *Hexaemeron* (v.12.39), takes its name from the fact that by its song it signifies *lucinia,* that is, the day bringing "light". While brooding its eggs it relieves the sleepless tedium of the long night by the sweetness of its song (H iii.33). PB (II,159) is the only French bestiary to mention the nightingale, which sings, according to the text, so enthusiastically at dawn that it almost dies.

Pliny has an unusually appreciative and lyrical account, for him, of the nightingale; he says that they sometimes vie with one another in song, and the one which is defeated often dies, abandoning life rather than its song (x.29.43).

There is little remarkable about the illustrations of the nightingale in Latin bestiaries except for B.M., Sloane 3544, f. 30, where a singing bird sits beside two sleeping men while a third stretches out his arm (imploring the bird to be silent?). In Montpellier H. 437, f. 208 (PB) a vanquished nightingale drops dead from one tree while the victor sits upon another.

ONAGER.

onager; *asne salvage.*

The early Latin Y version contains two chapters on the Onager: one on the Wild Ass alone; the other on the Onager and the Ape. The influence of this connection, although the ape's story is totally different, is still seen in the order of animals in PT, GC, and PB, where the chapter on the Ape follows that on the Onager. Y (11), after quoting Job 39:5, "Quis 'est qui dimittit' onagrum liberum", recounts how the male onager is first in the flock. If a male offspring is born, the father emasculates it lest it reproduce. In Y (25) the onager's announcing the arrival of the equinox is told. On the twenty-fifth day of the month of Famenoth, that is March, the wild ass brays twelve times to indicate the equinox. The ape urinates seven times. B (21)[105] differs slightly. The ass brays at the equinox twelve times in the night and in the day, and the hour is known by the number of brays. The ape is here said to have the appearance of the devil.

The onager symbolizes the devil who, when he sees the people who had wandered in the shadow of death converted to God and co-equal in the faith of patriarchs and prophets, brays, seeking the food he has lost. PT adds that the days stand for the good people, the nights the bad, the hours the number of people, and the equinox everlasting heaven and hell.

Following Isidore (xii.1.39), H (ii.11) and B-Is (21) tell of the equinox and the male's jealousy of male offspring which causes him to bite off their organs, although the mother has hidden her young (the last trait is omitted in B.M., Royal 2 C. xii, but is found in other manuscripts of B-Is).

Pliny (viii.30.46) speaks of the onager's fearing rivals in its lust, but the animal's link with the equinox appears much older. The baboon's relationship with the equinox is noted in Horapollo (i.16). In an Egyptian papyrus in the British Museum[106] Professor Perry has indicated a source for the substitution of the onager for the baboon. The ape's urinating might have arisen, according to Wellmann, through a confusion with the mystical nature of the cat.[107]

PT (1827-1850) speaks merely of the onager in connection with the equinox, whereas GC (1843-1880) is the only French writer to include the emasculation. PB (III,224) says that the ass cries out at the equinox, but he more fully develops the point that this animal makes the greatest effort to bray with its ugly voice, and that it only brays when it can find nothing to eat, and so strongly then that it almost bursts.

The Latin bestiaries are almost evenly divided in portraying a single braying ass and the emasculation scene, whereas GC more often depicts the latter. An unusually full picture of the braying, the hiding of the young, and the emasculation is found in B.N., fr. 14969, f. 35. An original manner of presenting the conception of the equinox to the eye is in the pair of scales beside the braying ass in a manuscript of DC (Pl. VII, Fig. 2).

---

[105] Professor Carmody includes the Onager and the Ape in the same chapter in his edition of *Versio B*, although they are almost without exception given separate headings in manuscripts belonging to the B and B-Is versions.

[106] PW, *op. cit.*, p. 1095.

[107] Wellmann, *op. cit.*, p. 66.

OSTRICH.

assida, struthiocamelon, struthio; assidam, ostrice, chamoi.

B (27) begins by giving the Hebrew, Greek, and Latin names for
the ostrich, and then quotes a pre-Vulgate reading of Jeremiah 8:7:
"Et 'asida' in celo cognovit tempus suum."[108] This bird is like a
vulture. It has wings but cannot fly, and it has feet like a camel's;
hence its name *struthiocamelon* in Greek. When it is time for laying
eggs, it looks to the sky and sees the star Virgilia rising,[109] that
is, around June. The ostrich digs in the earth, lays its eggs, covers
them with sand, leaves them and forgets them. The eggs are then
incubated by the warm summer sand.

Allegorically interpreted, this means that even more than the
ostrich man should forget the world in order to concentrate on
heaven. He should raise the eyes of his heart and love God more
than earthly ties.

It is probable that this story arose from combining the verse
cited from Jeremiah with Job 39:13-16 telling of the ostrich forgetting its eggs in the dust. The origin of the tale rests in part upon
the apparent nesting habits of this bird.[110] The description of the
ostrich as having wings useless for flying and a foot provided with
two toes (not, however, like a camel's) is mentioned by Aristotle
(*De part. anim.* iv 697b 13), although he calls the foot a cloven
hoof.

In the *Aviarium* (i.37) attention is given to the bird's inability
to fly, while the matter of leaving the eggs is developed only in
the moralization of the verses from Job.

No change takes place in the early French bestiaries, but through
carelessness PB contains two chapters on the ostrich: one (III,257)
entitled *assida*; the other (II,197) *ostrische*. The former follows the
Latin more closely but attributes the following description to Jeremiah: the ostrich has a neck and head like a swan, short fat legs,
cloven feet like a cow, and a body and tail like a crane. In the

---

[108] The Vulgate reads *milvus*, "kite", and the A.V. translates the word
as "stork".

[109] The Pleiades.

[110] Hastings, *op. cit.*, III, 635. See also Max Goldstaub, "Physiologus -
Fabelein über das Brüten des Vogels Strauss", *Festschrift Adolf Tobler*
(Braunschweig, 1905), pp. 153-90.

latter chapter the feet are said to be cloven like a stag's and the legs resemble those of an ass. *Phisiologe* is quoted as saying that the bird eats iron.[111]

The only way by which the ostrich in most cases is distinguished from other birds is its thick legs and cloven foot. The customary scene shows the eggs in sand while the bird looks up at a star. For its metal meal it eats a horseshoe or nails (Pl. VI, Fig. 3).

O W L.

nycticorax, noctua, bubo; *fresaie, huerans.*

In the oldest Latin versions (Y 7, B 7) the chapter on the Owl aways follows the Pelican's chapter because of the following verse from Psalm 101:7 (Vulgate): "Similis factus sum pellicano solitudinis: factus sum sicut nycticorax in domicilio." After citing the appropriate second clause, the versions conclude, in one of the shortest of all *Physiologus* descriptions, that the owl is a dirty bird,[112] preferring darkness to light. The owl's preference for darkness signifies the rejection of Christ by the Jews when they said: "We have no other king but Caesar." Christ then brought his light to the Gentiles. PT adds that the Jews are backward like the bird's flight.

The *Aviarium* (i.34) states that the owl dwells in roofless ruins and flies about at night seeking food. The same source (i.44) on the *bubo* cites Isidore as its model (xii.7.39). Here the owl is depicted as a heavily feathered, slothful bird which always hovers about tombs and dwells in caves. It is called *bubo* from the sound of its voice. This filthy bird pollutes its dwelling place with dung. When seen by other birds, it is betrayed by their clamoring and molested by their attacks. CUL's brief account differs slightly from the above and begins with Isidore's explanation that the *noctua* is thus called because it flies at night (*nox*).

PT (2789-2802) describes the *fresaie* as a dirty nocturnal bird which flies backwards. In PB (III,234) the bird called *huerans* likes

---

[111] This popular notion developed perhaps from a statement in Pliny (x.1.1) about the remarkable ability of the ostrich to digest anything.

[112] Deut. 14:15.

to frequent tombs, and when someone is near death, this owl senses it from afar and cries out.[113]

Owlish characteristics are sometimes absent from pictures of this bird, which is usually portrayed alone. However, in Bodl. 764, f. 73v. three small birds attack a strong featured *bubo*.

O X.

bos, urus, bubalus.

In Latin the ox is called *triones* because it "treads" or "rubs" (*terit*) the *terra*, "earth". The ox greatly enjoys company and seeks its accustomed partner to pull the plow, roaring frequently if separated. Oxen can predict the weather, and they remain in their stalls when rain threatens. The urus is a wild ox with huge horns which are used as drinking vessels on royal tables. In India there is a one-horned ox, extremely cruel and untamable, which has a solid hoof (H iii.19).

This relationship of *triones* and *terra* is found in Varro (vii.74) and then in Isidore (xii.1.30); it is Ambrose (*Hexaemeron* vi.4.20) who calls the oxen rain prophets. Barbarians of the North use the horns of the urus for cups according to Pliny (xi.37.45), who again mentions the urus when he says the common people in their ignorance call it bubalus (viii.15.15).

An unmistakable and uninteresting ox is always drawn.

PANTHER.

panthera; pantere, love cervere.

Y (29) begins with a pre-Vulgate quotation from Hosea (5 : 14): "Factus sum 'sicut leo domui' Iuda, et sicut 'panthera' domui Effraim." B (23), omitting this verse, says that the panther is a many-colored animal, extremely handsome and gentle, whose only enemy is the dragon. When the panther has eaten plentifully, it retreats to its cave and there sleeps for three days. Upon awakening it utters

---

[113] A second chapter on the Owl in PB (II,169) is entitled "le cauve soris". Either because of a strange coincidence or because of a definite relationship, the illustration in Queen Mary's Psalter (B.M., Royal 2 B. vii, f. 91v.) which in the normal order of contents would picture the owl shows instead a man beating off a bat.

a great roar and at the same time a sweet odor comes from its mouth. Animals from far and near hear the voice and follow the sweet breath except for the dragon, who from fear enters its hole where it remains stiff as if dead, lest the odor strike it.

In the longest of all *Physiologus* allegories the panther symbolizes Christ, who by his incarnation drew mankind to him, and the dragon represents the devil. The variegated appearance of the panther signifies the many qualities of Christ who, satiated with the mockeries of the Jews, died and was buried. He descended into hell, bound the great dragon, and arose on the third day. The panther's sweet breath is Christ's voice calling out after his resurrection.

H (ii.23), B-Is (24), who includes here a description of the dragon, and Isidore (xii.2.8,9) say that the panther is called $\pi\alpha\nu\theta\acute{\eta}\rho\alpha$ because it is a friend of "all animals" except the dragon. It is further described as having a tawny coat spotted with black or white disks. The female can only bear once because when the young in the womb reach the time of birth, they tear their mother with their nails so that great pain forces her to give birth. She is unable thereafter to conceive. Pliny is quoted as saying that frequently animals with sharp claws cannot have offspring because they are injured internally by the moving cubs.

The sweet odor of the panther is noted by Aristotle (ix 612a 13) but according to him it is used as a lure to catch animals.[114] This is repeated by Pliny (viii.17.23) and Plutarch (*De soll. anim.* 24 976D) among others. Isidore's allusion to Pliny is in reality the latter's reference to the lioness who at delivery tears her womb with her claws (viii.16.17).

None of the French versions repeats the account of the birth of the panther's young, but all mention the attraction of its breath and its enmity toward the dragon. GC (2029-2068) says that the animal whose name is panther is "en dreit romanz love cervere", "lynx". Its coloring seems to have had a special fascination for this author.

> Car ele est blanche e ynde e bleue
> E jalne e verte e russe e bise
> E coloree en meinte guise.   (2034-2036)

---

[114] Sbordone has pointed out the change effected in the ancient tradition of the mortal attraction of the panther's breath when, in its Christian form, the panther became the symbol of Christ (Sbordone, *Ricerche*, p. 104).

The miniatures of the panther are always recognizable, not because of the appearance of the principal animal, which varies from that of a dog to a horse, but because of the variety of animals clustered about it, and the serpent's tail disappearing into a hole. An amusing marginal addition is seen in Bodl. 602, f. 21v. The large illustration (Pl. VI, Fig. 4) is a typical representation of the panther and beasts, but the marginal drawing on the following folio portrays the panther seated on its haunches holding in its paws the enemy serpent toward which it spits its pleasant odor. In two related manuscripts of GC (Camb., Fitzwilliam McLean 123, f. 49v., and Bodl. 912, f. 7v.) the panther is an odd looking beast-serpent with two feet, wings, and a curved tail. Among the congregated animals is a camel!

PARANDRUS.

parandrus, tharandus.

The parandrus, which comes from Ethiopia, is the size of an ox and the color of a bear, has cloven hoofs,[115] branching horns, a stag's head, and long hair. It has the habit of changing its appearance so that when concealed it becomes like its surroundings (H iii.9; Solinus 30.25).

Pliny (viii.34.52) mentions that the *tarandrus* of the Scythians changes color, and from his description the animal has been variously identified as the reindeer or the elk.

The parandrus in Morgan 81, f. 39 is an abundantly tufted animal with magnificent antlers.

PARD and LEOPARD.

pardus, leopardus.

The pard is a kind of spotted wild animal, extremely swift and bloodthirsty, for with one leap it strikes its victim dead. The leopard is born from the adultery of the lioness and the pard. Pliny

---

[115] The manuscript of H published by Migne reads *bisulco vestigio,* "cloven sole", as does Solinus, but CUL, f. 16 reads *ibico vestigio,* which White in *The Book of Beasts,* p. 52, translates as "the slot of an Ibex".

is quoted as speaking of another origin: the lion's mating with the female pard, from which a degenerate birth results (H iii.2; Isidore xii.2.10,11). Pliny (viii.16.17) mentions the male panther, calling it a pard, and its relations with the lioness.

An animal spotted with circles is drawn in Camb., Fitzwilliam Museum 254, f. 15v.

PARROT.

psittacus; *papegai*.

The parrot comes from India. It is a green bird with a purple-red collar and a tongue which is bigger and broader than that of other birds. It utters words so distinctly that if unseen, one would think it a man. By nature it greets, saying *ave* or χαῖρε, but other words it must learn by teaching. Whence Martial (xiv.73) says:

> Psitacus a vobis aliorum nomina discam;
> Hoc didici per me dicere cesar ave.[116]

Its beak is so hard that when it falls from on high upon a stone, it catches itself by the pressure of its mouth. The parrot learns more quickly and retains better when it is young (H iii.28; Isidore xii.7.24; Solinus 52.43).

Only PB (II,186) among the French works mentions the parrot. He states that there are two kinds: those which have three toes are of a mean disposition and those with six are gentle. The parrot is also said to hate rain because it makes its color ugly. Pliny (x.41.58) lists neither of these details in his description of the parrot.

Another of many examples showing the artist's lack of observation of birds or his failure to illustrate any statement in the text is evident in the drawing in CUL, f. 33, where the parrot is a long-billed bird standing on water. Evidently the bird is being taught and chastised in Camb., Gonville and Caius Coll. 384, f. 182v., where a man holding a stick kneels before a parrot which bears no resemblance to the actual bird.

---

[116] "I, a parrot, will learn the names of other things from you;
This I have learnt to say by myself: 'All hail, O Caesar'."
(Loeb trans.)

PARTRIDGE.

perdix; perdrix, pertris.

The partridge is a deceitful bird, so Y (31) and B (25) begin, then quote from Jeremiah (17:11): "Clamavit perdix et congregavit quae non peperit." This bird steals the eggs of others and broods them, but she can have no fruit from her deceit for when the young partridges hear the voice of their mother, they fly off to their true parents. The partridge signifies the devil, who steals God's offspring, but they, on hearing Christ's (or the church's) voice, fly off and are received under his wing and given to mother church.

The *Aviarium* (i.50) repeats this description and adds from Isidore (xii.7.63) that it takes its name from its voice. CUL includes several different details. Males occasionally have intercourse, and the female is so lustful that wind alone from a male can impregnate her. Partridges camouflage their nests with thorns and cover the eggs with dust, but should a man approach the nest, the mother feigns a broken foot or wing and draws the man away from the nest. When it has been discovered, the partridge lies on its back and covers itself with clods of earth.

There is no trace of the fraudulent partridge in Greek literature although Aelian (iii.30) writes a somewhat related story about the cuckoo, which places its eggs in others' nests. The Hebraic word for the "caller" or "crier" was translated by the Septuagint as πέρδριξ and thus it appeared in the passage from Jeremiah quoted above. In Jerome's commentary on this verse he writes:[117] "aiunt scriptores naturalis historiae tam bestiarum et volucram quam arborum herbarumque (quorum principes sunt apud Graecos Aristoteles et Theophrastus, apud nos Plinius Secundus [sic]) hanc perdicis esse naturam..." and tells of the theft of the eggs by a partridge in the traditional manner that Ambrose also recorded (*Hexaemeron* vi.3.13). Since none of the writers named by Jerome describes the partridge in this way, it appears that the story of the partridge was contrived to explain the verse of Jeremiah.[118] All of the additions in CUL are found in Pliny (x.33.51).

There is no difference in the French accounts of the partridge. GC (2345-2380), however, following B-Is (26) is the only author

---

[117] Migne, *Patr. Lat.*, XXIV, Col. 820.
[118] Sbordone, *op. cit.*, p. 68.

who mentions the unnatural love of the males. He says too that in spite of their being unclean, they are eaten.

The illustrations of the partridge, like many of those depicting the birds mentioned in the bestiary, instead of clearly stressing the bird's principal trait, merely portray a standardized bird, or occasionally a large bird with some young, but with nothing except the text to identify it as the partridge and its tale. More vivid is the illustration in B.M., Sloane 3544, f. 29, where a bird walks with an egg in its beak from one nest to another, there to sit on it.

PEACOCK.

pavo ; *paon.*

The *Aviarium* (i.55) cites Isidore (xii.7.48) as its source for the derivation of the peacock's name from the sound of its voice, but it explains further that when the bird unexpectedly begins to cry, it strikes the listener with *pavor,* "fear". It has hard flesh, which resists putrefaction but which can scarcely be cooked by fire or even by the heat of the liver in the stomach. The voice of the peacock is terrible, its head serpentine, its breast sapphire colored. Small red feathers are in the wings, and the long green tail is adorned with eyes. When praised, it raises its tail, leaving its rear part bare. This version omits Isidore's citation of an epigram by Martial (xiii.70).

> Miraris quotiens gemmantes explicat alas.
> Et potes hunc saevo tradere dure coco.[119]

PB (II,161) says that the peacock suddenly awakens and cries out, thinking its beauty lost. The prophet Amos is credited with the statement that it is a bird of great foresight.[120]

---

[119] "You admire as often as it spreads its jeweled wings,—
And then hand this creature over to a cruel cook,—
you hard hearted man."　　　(Loeb trans.)

[120] None of the traditional Latin bestiaries with Isidore's additions recounts a trait that is often used later to counteract the peacock's pride—the bird wishes to fly, but at the sight of its ugly feet, rather than fly high it remains grieving. This description is found in the Waldensian Bestiary (4). The peacock's screaming at the sight of its feet is mentioned in the *Physiologus* attributed to Epiphanius (Migne, *Patr. Gr.,* XLIII, Col. 527).

From Varro (v.75) is taken the statement that the peacock's name comes from its voice. Saint Augustine, testing the ability of the peacock's flesh to withstand decay, discovered that after a year's time the cooked meat was only somewhat shriveled and dried! (*De civitate Dei* xxi.4). Aelian (iii.42) notes that the bird is adorned with feathers, but the rest of its body is not seemly.

The general effect of the peacock is well portrayed in the bestiaries with a magnificent tail either lowered as in Morgan 81, f. 56 or a frontal view of the bird with its tail outspread as in Bodl. 764, f. 84v. The only illustration seen where a noticeable attempt to give the peacock unattractive feet is in Arsenal 3516, f. 202 of PB, where the text does not refer to them.

PEARL and AGATE.

margarita, unio, perla, concha, concha sabea, mermecolion, achates; union.

Although Y (22) has a separate chapter on the Agate, it is here treated with the Pearl because it occasionally so appears in the earliest accounts (C 23) and then ceases to exist,[121] while the Pearl continues its existence in the Latin and in one French bestiary.

To find pearls divers tie an agate to a rope which is dropped into the sea. The stone comes to a pearl, remains there, and the diver follows the rope to its treasure.

According to Y (23) before dawn at sea the stone which is called oyster (*sostoros*)[122] opens its mouth and swallows dew, the rays of the sun, moon, and stars. From this the pearl is born.[123]

The essential part of the long allegory as recorded in B-Is (37) is found at the beginning, which says that the pearl signifies the Virgin Mary, who ascended to the temple of God and there received the words (celestial dew) of Gabriel. The opening of the shell symbolizes the Virgin who said "Ecce ancilla Domini...". PT adds that as the shell opens and closes without a break, so did the Virgin conceive and give birth.

---

[121] Unexpectedly the account of the Pearl and Agate appears in H (iii.57) after the long section on Trees, and in B.M., Harl. 3244, f. 71v.

[122] In C (Bern 318) the *conchos* is said to be a fish.

[123] Professor Carmody does not print the text on the Pearl in his *Versio B*, but sends the reader to B-Is (37).

Isidore's account (xii.7.49) is followed in some later bestiaries which call the pearl *oceloe*. This word has numerous spellings and its origin is somewhat uncertain.[124] In H (ii.35) pearls are called *uniones*, though the common people say *perlae*. Of these a certain kind are called *marmaetholion* (*mermecolion*), for which the Greek word is *concha sabea*. In manuscripts of the common B or B-Is version this passage begins "Item lapis est in mari qui dicitur latine mermecolion, grece concha sabea, quia concavus est et rotundus". What *concha sabea* means or why the name that was attached to the Ant-Lion, *mermecolion*, is found also applied to the pearl remains so far unexplained.[125]

The role of the agate in finding pearls is unknown before the *Physiologus*, but in Arrian's *Indica* (viii.8) Megasthenes reports that should the king pearl be captured, the others are easily caught. The birth of the pearl from dew is recorded in classical Indian poetry.[126]

Two descriptions of the pearl are found in PT (3015-3062), which include some statements similar to those in Pliny (ix.35.54,56), where the pearl is called *unio* as in PT and where the island of Taprobane (Ceylon, PT *Tapne*) is said to be very fertile in pearls. PT adds that if one drinks the pearl mixed with dew it will cure any illness but death.

The only miniature seen of the agate's use in finding pearls is Bern 318, f. 20v., which shows two men in a boat while a third dives into the water, guiding himself with a rope. In Bodl. 602, f. 35 the two valves of an oyster are open to receive drops of dew from the sky and rays from the sun. Beside a closed oyster, to illustrate the allegory, is a graceful drawing of the Virgin holding her young Son.

PELICAN.

pelicanus, onocrotalus; *pellican*.

B (6) begins by quoting Psalm 101:7 (Vulgate): "Similis factus sum pelicano solitudinis." The pelican is said to have great love

---

[124] Cf. Lindsay's edition of Isidore. A passage very similar to Isidore's in Rabanus Maurus (Migne. *Patr. Lat.*, CXI, Col. 238) probably offers an explanation: "...ex coelesti rore margaretum concipiunt. Unde et coeloe nominantur."

[125] James, *op. cit.*, pp. 8-9.

[126] Sbordone, *op. cit.*, p. 134.

for its young. However, when these begin to grow, they strike their
parents in the face. Their parents in turn strike and kill them. After
three days their mother pierces her side and sheds her blood over
the dead children, thus reviving them.

Allegorically Christ is the pelican whom mankind struck by
serving what has been created rather than the creator. Christ then
ascended the cross, where from his pierced side flowed the blood
and water of man's salvation and eternal life.

In Horapollo (i.11) the vulture is spoken of as feeding its young
on its own blood when unable to find food. Wellmann believes it
plausible that the name ῥάμφος, "crooked bill", as the pelican is
called in Hermes' *Koiraniden,* was also applied to the vulture, hence
the transfer of the story.[127] Sbordone, on the other hand, is of the
opinion that the tale of resurrecting the dead with one's own blood
could only be the work of a Christian writer.[128]

H (ii.27) and B-Is call the pelican an Egyptian bird living in the
Nile (Isidore xii.7.26 adds that is from this country that the pelican
gets its name, for Egypt is called Canopos). There are two kinds
of pelicans: one lives in water and feeds on poisonous animals
such as lizards and crocodiles; the other, the onocrotalus, a bird
with a long neck and bill, imitates the braying of an ass when it
drinks. The account in the *Aviarium* (i.33) offers an explanation
of its eating habits different from the above and apparently derived
from Pliny (x.47.66). Whatever this insatiable bird swallows is immediately digested because its stomach has no diverticulum in which
to retain food.

PT (2323-2366), following the order of B-Is, has reversed the
two types of pelicans. According to him those living in the water
eat fish, and those living on islands eat dirty animals. He gives
honocrotalia as the Greek name,[129] explaining that in Latin this
is *longum rostrum* and in French *lunc bec.* Here also there is a
variation which is found in some Latin bestiaries and is carried
through their French translations. Instead of the mother's resuscitat-
ing the young, it is the father who, regretting his action which had
caused their death, pierces his side.

The two scenes of the mortal clawing between two birds and
the pelican piercing its side, from which blood streams down on

---

[127] Wellmann, *op. cit.,* p. 50, n. 142.

[128] Sbordone, *op. cit.,* p. 16.

[129] The word *onocrotalus,* "ass-clapper", is applied to the pelican by
Pliny (x.47.66).

the young, are often simultaneously presented (Pl. VII, Fig. 4), although in several illustrations there is no reviving blood, but only the fighting among the birds. An unusual depiction of unknown origin is that in Morgan 81, f. 6v., where the parent's blood flows from its mouth to the open beak of a smaller bird.*

PERIDEXION TREE.

peredixion, circa dexteram, pendens, perindens;[130] *paradixion, environ destre.*

There is a tree in India which is called in Greek *peredixion* and in Latin *circa dexteram*. Its fruit is so sweet that doves like to sit in the tree. A cruel dragon hates the doves and they are afraid of him, but the dragon fears the tree's shadow even more and flees it. When the shadow of the peridexion is on the right, the dragon moves to the left, and the opposite is also true. The doves know of the timidity on the part of the dragon and stay safely in the shadow of the tree, but should a dove leave the shadow, it would be devoured by the dragon (Y 19; B 32; B-Is 33; H iii.39). The French versions are the same.

In the allegory the doves are the faithful, who are admonished to be both simple and shrewd and to remain within the church. Christ is portrayed as the right side of the tree, while the shadow is the Holy Ghost. The faithful are to beware lest the devil-dragon find them outside God's house.

In his book on trees Pliny (xvi.13.24) recounts that no serpent will ever lie in the shadow cast by the ash, though no mention is made of doves. Lauchert believed that the doves were added by the author of the *Physiologus* to the story of the peridexion tree under the influence of the Biblical parable of the grain of mustard seed which became a tree "so that the birds of the air come and lodge in the branches thereof" (Matt. 13:32) or, as Mark says (4:32), "So that the fowls of the air may lodge under the shadow of it".[131] However, because of the similarity in wording Wellmann considers the source of the *Physiologus* story to be one such as Hermes used in his version of the peridexion

---

[130] In Camb., Corpus Christi Coll. 22, f. 169, the peridexion tree is called *mandragora ... vel perdixion*.

[131] Lauchert, *op. cit.*, p. 29.

tree, which included doves in his *Koiraniden*.[132] Finally, it is Sbordone's opinion that the account was entirely a creation of the
author of the *Physiologus* and then transcribed by Hermes.[133]

Just as the accounts of the peridexion tree are unusually consistent, so the drawings show little variety. All contain a tree which is
usually an ornate, symmetrical stylization with doves perched on
its leaves and a winged dragon at its base. Exceptional is the
simplicity of Bodl. 602 (Pl. VII, Fig. 5). In the center of the tree
in Arsenal 3516, f. 209 (PB) sits the figure of God blessing the
doves on the tree's edge.

PHOENIX.

phoenix; fenix.

In India the phoenix at the age of five hundred years goes to
the frankincense tree ("intrat in lignis libani") and fills its wings
with spices. In the month of March or April[134] a priest of Heliopolis
piles twigs on an altar. When the bird arrives in the city and sees
the altar, it rolls in spices, lights a fire, and is consumed. The
following day the priest finds on the altar a small, sweet-smelling
worm. On the second day it has the appearance of a bird, and on
the third day the priest discovers a perfect phoenix.[135] The bird
bids the priest farewell and returns to its place of origin (B 9; Y 9).
All versions agree that the phoenix symbolizes Christ, who had
the power to come back to life.

Many Latin versions of the Second Family, such as CUL, contain
two accounts which differ slightly from each other. The first of these
resembles the description in the *Aviarium* (i.49) and the second
copies the notice in Ambrose's *Hexaemeron* (v.23.79). The *Aviarium*
text, identical with Isidore (xii.7.22), omits mention of the priest
and Heliopolis, and calls the phoenix a bird of Arabia either because
of its purple red color (*phoeniceus*) or because it is unique in the
world and the Arabs by *phoenicem* mean a "single thing". Once
on the pyre which it has built, the bird turns toward the sun and
fans the fire by the beating of its wings so that a second time it

---

[132] Wellmann, *op. cit.*, pp. 51-52.
[133] Sbordone, *op. cit.*, p. 109, and *Ricerche*, pp. 84-89.
[134] Both B and Y give the Coptic names for these months.
[135] Y declares that it is a large eagle which is found on the third day.

might arise from the ashes. B-Is, as would be expected, contains B's version and Isidore's, making again two separate accounts only one of which is similar to the Second Family.

Tales of the burning and rebirth of the phoenix are among the oldest and most widely diffused of antiquity. For the Egyptians the phoenix was a symbol of the rising sun and thus linked with the idea of a resurrection.[136] In various forms, none closely approximating the *Physiologus* versions, the story of this bird is found in Herodotus (ii.73), Pliny (x.2.2), and Aelian (vi.58). Only later in the writings of Tertullian and Clement of Rome does the account approach B and Y.[137]

There are only small changes in the Old French versions of the phoenix. PT (2217-2203) gives the two successive accounts found in B-Is. In the first, from Isidore, he says that the bird is wholly purple in color and formed like a swan;[138] the second description follows B. According to GC (739-786), instead of the wings catching fire from the sun's rays as in PT, the fire is lit by the striking of the bird's beak against a stone. G (1021-1046) says that along with the spices the phoenix gathers stones, which cause the fire that is fanned by the bird's wings. How the fire-kindling properties of the stone entered the phoenix story, I do not know. As usual PB (II, 182) gives a colorful description, probably from Solinus (33.11). The bird

---

[136] See Cook, *op. cit.*, pp. lxxv-lvi, for a detailed account of the symbolism of the phoenix, and for various identifications of the bird see *Lactanti De Ave Phoenice*, ed., Mary C. Fitzpatrick (Published Ph.D. dissertation, University of Pennsylvania, 1933), p. 18, n. 29. A further study is that by Jean Hubaux and Maxime Leroy, *Le Mythe du phénix dans les littératures grecque et latine* (Bibliothèque de la Faculté de Philosophie et Lettres de l'Université de Liège. Fascicule LXXXII. Liège, 1939).

[137] Tertullian, *Liber de resurrectione carnis* (Migne, *Patr. Lat.*, II, Col. 857), connects the phoenix bird with the Septuagint rendering of Psalm 91:13: "Et florebit velut phoenix", where the Greek word θοῖνιξ meant "palm tree". The account of Clement of Rome, *Epistola I ad Corinthios* (Migne, *Patr. Gr.*, I, Cols. 261-265, is somewhat similar to Ambrose's description.

[138] PT in describing the phoenix as

> En Arabie est truvez,
> Cume cisne est furmez;   (2219-2220)

must have hastily read Isidore's introduction, "Phoenix Arabiae avis, dicta quod colorem phoeniceum habeat", and misunderstood*colorem*, "color", for *olorem*,"swan".

is crested like a peacock, its chest and throat are resplendent with red and shine like pure gold, and towards the tail it is sky blue. When it flies to a mountain called *Liban,* it finds a high tree by an excellent fountain. After that the text follows B.

The earliest Latin bestiary drawing of the phoenix is that of Brussels 10074, f. 145 where the bird sits in a tree labeled *cedrus libansi.* Beside this scene a priest is pictured finding the new bird upon an altar. An attendant brings faggots to the burning phoenix in Bodl., Douce 88 (Pl. VII, Fig. 1). The priest and the reborn bird appear in two manuscripts of GC (B.N., fr. 14969, f. 14v. and B.M., Cott. Vesp. A. vii, f. 9). The most common pose, however, is the phoenix plucking twigs to make its pyre or sitting upon the burning nest.

QUAIL.

coturnix.

Quails are so called from the sound of their voice, but the Greeks call them *ortygas* because they were first seen on the island of Ortygia.[139] At summer's end they migrate over the seas. The bird which leads the flock is called *ortygometra,* "mother of quails". When this bird is close to earth, it is snatched by the hawk, and for this reason quails seek a leader from another species. The quail alone, like man, suffers from falling sickness or epilepsy (*Aviarium* i.51; Isidore xii.7.64,65). CUL adds, from Isidore, that quails feed on poisonous seeds, and for this reason the ancients forbade their being eaten.

It is Festus who is recorded by Paulus as giving the derivation of the bird's name from its voice. The remainder of the account is found in Pliny (x.23.33) and in Solinus (11.20), who explains that the Greeks believed the quail to be under the guardianship of Latona, the goddess who gave birth to Artemis and Apollo on Delos.

The hawk attacking the quails' leader and holding it by the neck is depicted in Bodl. 764, f. 82v., but in most manuscripts the quail is a very nondescript bird.

---

[139] Ortygia, or Quail-Island, is the ancient name of Delos.

RAVEN.

corvus, corax; *corbeau*.

It is easier to read CUL's description of the raven, which is copied from Isidore (xii.7.43), than to extract it from that of the *Aviarium* (i.35), where it is mingled with Gregory's commentary on Job 38:41: "Quis praeparat corvo escam suam, quando pulli eius clamant ad Deum vagantes, eo quod non habeant cibos?" Isidore says that the raven, *corvus* or *corax*, takes its name from the sound of its wind-pipe because it "croaks" (*coracinet*). It does not bring food to its young until it can recognize them by the black color of their feathers. In a corpse the raven first seeks the eyes.[140]

The raven enters French bestiaries in G, who speaks only of the extreme caution of the parents in feeding their offspring. The featherless young in PB (II,156) are fed by dew until their feathers grow and cause them to resemble their father. He also explains that the raven eats human eyes in order to be able to draw forth the brain.

Drawings of the raven's activities are usually more vivid in French manuscripts than in Latin. Dew rains into the open beaks of the young birds in Arsenal 3516, f. 201 (PB) as the raven pecks the eyes in a corpse's head.

SALAMANDER.

salamandra, stellio; *salamandre, sylio*.

According to B (30) there is a reptile which is called *salamandra* by the Greeks and *stellio* by the Latins. Concerning it Solomon said: "Sicut stellio 'habitans in domibus' regum" (Prov. 30:28).[141] It is like a small multi-colored lizard. If it should come upon a fire or a burning furnace, the fire is immediately extinguished. Y (45) is similar to this, but C (18) exhibits an interesting difference that

---

[140] The probable origin of this trait is found in Proverbs 30:17: "The eye that mocketh at his father, and despiseth to obey his mother, the ravens of the valley shall pick it out, and the young eagles shall eat it."

[141] The A.V. reads: "The spider taketh hold with her hands, and is in the kings' palaces."

exists in Sbordone's first Greek text: not only does the salamander extinguish fire, but should it enter a bath, the water would become cold.

The salamander signifies just men, like the three that Daniel spoke of who emerged unharmed from the fiery furnace, because the faithful can penetrate fire with impunity.

Largely copying Isidore (xii.4.36), H (ii.16) and B-Is (31) say that the salamander's venom is very powerful for where others kill singly, it kills many at once. It poisons apples on a tree and water in a well. This animal not only lives painlessly in flames, but extinguishes them.

Even Aristotle believed that the salamander was an animal which fire could not destroy (v 552b 16). The salamander's poison is mentioned by Pliny (xxix.4.23), who also attributes the extinguishing of fire to the coldness of the animal (x.67.86). This is perhaps the basis for the action found in C of making water frigid.

In the French bestiaries the salamander starts out as normally as such a reptile can, with its own chapter, but later it is incorporated into the section on the four Elements as representing fire, and eventually it becomes a bird in the *Bestiaire d'Amour*. PT and GC repeat B-Is, but a new trait is added in PB (III,271). From the salamander, which lives on pure fire, is produced something that is neither silk, linen, nor wool, but another material which is worn by people of importance. When dirty, the only means of cleaning this cloth is to put it in fire, where it does not burn. In his chapter on the Elements (IV,77) PB declares that the salamander is said to have fleece like a lamb's "mais nus ne peut savoir quel cose ce est". Cloth is made from this unknown substance in a part of the deserts of India.

The portrayal of the salamander is even more varied than its lizard-to-bird descriptions, and its different qualities are sometimes combined in a very compact fashion. The oldest illustration is that in Bern 318, f. 17v., where the salamander is a satyr-like creature in a circular wooden tub. Following the text, this must be the salamander cooling the bath water. The poisonous effects of the reptile are seen in B.M., Royal 12 C. xix (Pl. VII, Fig. 3). There are also less imaginative renderings of the salamander as merely a worm penetrating flames in Bodl. 764, f. 55, but GC, B.N., fr. 1444, f. 253v. portrays a winged dog and B.N., fr. 14970, f. 23v. depicts a small bird in flames.

SAWFISH.

serra; *serre, sarce.*

There is a monster in the sea, called a *serra*, which has enormous ·
wings. On seeing a ship it raises these wings and races against the
ship for thirty or forty stadia. Growing tired, it lowers the wings
and the waves carry the sawfish back to its original location in the
depths (Y 4; B 4). In DC (10), instead of wings the sawfish has
spinas ("spinas habens proppe se longiores"), although this probably originally was meant for *pennas* since the text continues by
saying that it lifts its wings.

Allegorically the sea represents the world in which the righteous
people or the apostles sail, while the sawfish signifies those who
have started out with good intentions but abandon them and revert
to sin. For PT the sawfish symbolizes the devil, who holds back
holy inspiration from men and captures them just as the *serra* draws
the wind away and devours fish.

H (ii.22) and B-Is repeat the older Latin versions without making any reference to Isidore's description (xii.6.16) of the *serra* as
having a serrated crest ("serrata crista") and swimming under ships
to cut them.[142]

The tale of the sawfish recalls Pliny on the swift dolphins which
compete with ships (ix.8.7,8). In Aristotle (ix 631a 21 ff.) the dolphin,
after pursuing fish, can hold its breath and shoot up like an arrow,
sailing over a ship's mast if one is nearby. Isidore's description seems
to be based on Pliny's reference to the sword-fish (xxxii.2.6) as
having a sharp-pointed muzzle with which it can pierce the sides
of ships and sink them.[143]

---

[142] In most Second Family manuscripts, where the *serra* is included in
the section on Fish, there is no illustration of this sea creature. In CUL
there are in reality two notices on the Sawfish: one following the paragraph
on *balene*, "whales", which concludes the B text on the Aspidochelone;
and the other following the short description of the *gladius*, "sword fish",
where the text copies Isidore.

[143] The text for the *serra* in Morgan 81, f. 68, is unique. Translated it
reads: "The city of Syria which is now called Tyre, was formerly named
Serra from a certain fish which used to abound there. And this fish they
called in their tongue "sar", from which it was deduced that little fish of
similar appearance to it were called sards or sardines." This description
comes from Isidore (xii.6.38), but the scribe has written *Serra* instead of
Isidore's *Sarra*, the old name of Tyre. Presumably the scribe confused the

An odd description occurs in PT (1681-1702).

> Serra beste est de mer,
> Eles at pur voler
> E teste at de leün
> E cue at de peissun.   (1681-1684)

When the sawfish sees a ship, it raises its wings to do great harm, because it flies ahead and holds back the wind from the sails. When it cannot go faster than the ship (though from what precedes, one would think the ship becalmed), it plunges into the sea to devour fish. The stopping of the ship is apparently related to Isidore's account (xii.6.34) of the *echineis* or *remora*, which clings to ships and renders them incapable of moving. GC (399-420) continues the notion of the harmful effect of this creature by saying that sailors dislike it because it often causes ships to sink. G (1105-1122), differing from DC, calls the *sarce* a crested bird with a wide notched tail. According to PB (II,121), although the beast has wings and flies, it swims against ships ("ele ... se lance parmi la mer et commence a nagier contre la nef ...").

The depiction of the sawfish is probably the most varied of all the fantastic animals in the bestiary — it ranges from a winged dog to a feathered fish and from harpy-like creatures to sirens. Sometimes influenced by the text and again uninhibited by it, the artist gave his imagination free line with the sawfish. Although each drawing has some peculiarity that should be mentioned, only a few can be indicated here, beginning with the oldest, Bern 318, f. 18v. Though the picture is difficult to make out, two men appear to be in a boat, and one of them seems to beat off with an oar a long fish with dorsal fins which swims abreast of the ship.[144] A startling phenomenon takes place in the second oldest manuscript and is perpetuated in the winged siren in Camb., Univ. Lib. Gg.6.5, f. 95. In Brussels 10074 (Pl. VIII, Fig. 1 b) the drawing for the text on the *serra* shows a siren beside a boat of sleeping men. On each of her arms and hands are attached five wings. Druce accepts Cahier's suggestion that the artist interpreted "pennas immanes"

---

two names, and copied this notice rather than xii.6.16. See George C. Druce, "The Legend of the Serra or Saw-Fish", *Proceedings of the Society of Antiquaries of London*, 2nd. Series, XXXI (1919), 28.

[144] The text of this manuscript begins "Haec piscis longas habet alas".

as equivalent to "pennas in manibus".[145] An illustration for DC (Morgan 832, f. 5v.) presents an empty boat beside which stands in the water a two-legged, animal-faced creature with wings and a wide tail, while another DC manuscript (B.M., Sloane 278, f. 51) contains a strange, web-footed bird with a cock's head, doubtless the influence of the word *crista,* "crest" or "cock's comb". The Copenhagen manuscript of PT (Pl. VIII, Fig. 1 a) shows the swallowing of fish. Numerous other variants could be given, such as the winged fish on Morgan 81 (Pl. VII, Fig. 6), but to end on a note of restraint, B.N., fr. 14970, f. 4 of GC pictures the sawfish as a small flying bird, and Bodl., Douce 132, f. 65, as a huge fish with two sets of wings.

SCITALIS.

scitalis, scytale.

H's chapter on the scitalis (iii.43) is a copy of Isidore (xii.4.19), who in turn repeated Solinus (27.29) with one addition. The scitalis is thus called because its back gleams so that the pleasure of beholding its markings slows a person down.[146] Those that it cannot pursue are caught, stunned by the marvelous appearance of the scitalis. So great is this serpent's heat that even in winter it sheds[147] its skin, whence Lucan (*Pharsalia* ix.717) wrote:

> Et scitalis sparsis etiam nunc sola pruinis,
> Exuvias positura suas.[148]

In Camb., Gonville and Caius Coll. 384, f. 98 the drawing shows that some attention has been given to the text because the snake-like creature has round marks down its back. It is usually portrayed as an undistinguished winged serpent.

---

[145] Druce, "The Legend of the Serra", *op. cit.,* p. 28.

[146] Actually the serpent's name comes from the Greek σκυτάλη, "staff", referring to its uniform thickness, according to the *NED.*

[147] CUL reads here instead of *deponat, exponat,* "displays".

[148] "The scytale, which alone can shed its skin while the rime is still scattered over the ground." (Loeb trans.)

SHEEP.

ovis.

The *ovis*, "sheep", is a defenseless, placid animal covered with soft wool whose name is derived from *oblatio*, "offering", since the ancients in the beginning offered sheep, not bulls, for sacrifice. Among these some are called *bidens*, either because two of their eight teeth are larger — these are the sheep that are sacrificed by the Gentiles —or because they are *biennis*, "two years old". In early winter they gather grass before it is nipped by frost (H iii.13; Isidore xii.1.9). The explanation of the name *bidens* seems to have been drawn from Servius' commentary on the *Aeneid* (iv.57).

The illustration of the sheep is always a faithful depiction of this animal.

SIREN and ONOCENTAUR.

sirena, onocentaurus; *serena, sereine, onoscentaurus.*

These two mythological figures are almost always contained in the same or in successive chapters in Latin and French bestiaries. Usually the notice on the Siren and Onocentaur is followed by that on the Hedgehog because of its inclusion — a pre-Vulgate usage, it should be noted — in the verse from Isaiah (13:22) with which Y (15) and B (12) begin: "Sirena et 'daemonia' saltabunt in Babylonia, et 'herinacii' et onocentauri habitabunt in domibus eorum."[149]

Sirens are deadly creatures who have a human form from the head to the navel; thence to their feet they have the appearance of birds. Their beautiful singing charms men, and from a distance they attract sailors. The sailors fall asleep, enchanted by the sirens' song. Then the sirens attack the men and tear their flesh. The onocentaur also has two natures: his upper part is like a man; his lower part exceedingly wild, or as Y says, like an ass.[150]

---

[149] Jerome's translation (Migne, *Patr. Lat.*, XXVIII, Col. 842) reads: "Sed requiescent ibi bestiae, et replebuntur domus eorum draconibus, et habitabunt ibi struthiones, et pilosi saltabunt ibi, (22) et respondebunt ibi uluae in aedibus eius, et sirenae in delubris voluptatis."

[150] For a summary of the early history of the onocentaur, see Robin, *op. cit.*, pp. 81-82.

The plight of the sailors shows that those who delight in the pleasures of this world are prey for the devil. PT gives more details in his moralization where the siren symbolizes wealth in this world which causes man to sin. The onocentaur is like two-tongued hypocrites who speak of doing good but act in an evil manner. PT differs slightly. Man is rightly called man when he tells the truth, but the ass signifies his evil deeds.

Isidore (xi.3.30), drawing in a certain measure upon Servius (*Comm. in Verg. Aen.* v.864), says that sirens are part women, part *volucres*, "birds", with wings and talons. With Hugo (ii.32) an important change has taken place; from their navel to their feet the sirens have the appearance of fish. B-Is continues Isidore's bird figure. Copying Isidore, this version states that one of the sirens sings, another plays the flute, and the third, a lyre. In truth, it is said, sirens are harlots who lead men into poverty and are therefore said to cause a shipwreck.

The siren was included in the *Physiologus* because of Jerome's use of the word *sirena* in the Vulgate. Homer, in the earliest reference to sirens (*Odyssey* xii.166 f.), does not describe them physically, but only alludes to the mortal charm of their voices. Ovid (*Metam.* v. 552), in the Greco-Oriental tradition, refers to their bird-like appearance:

... vobis, Acheloides, unde
Pluma pedesque avium, cum virginis ora geratis?

When then did the siren acquire the fish tail with which it is now associated? Edmond Faral has called attention to what he believes is the first mention of this new type of siren.[151] It is contained in the late seventh or early eighth century *Liber monstrorum*. At the beginning of the twelfth century the siren is still a woman-bird, but in the second quarter of the century the woman-fish appears in the vernacular, and the text of PT is the first to record this change of appearance.

PT (1361-1374) says that the siren sings in the tempest and weeps in fair weather. He describes her as being a woman to the

---

[151] Edmond Faral, "La queue de poisson des sirènes", *Romania*, LXXIV (1953), 433-506. Druce believes that the mermaid or woman with a fish tail has its origin in a classical source, that of the female triton (p. 174). See George C. Druce, "Some Abnormal and Composite Human Forms in English Church Architecture", *Archaeological Journal*, LXXII (1915), 135-86.

waist, having falcon's feet and the tail of a fish. The quotation
on the onocentaur (1109-1116) is attributed by PT to Isidore. GC
(1053-1070) compromises on the siren and declares that she can
be either a fish or a bird below the waist. What is now the current
description of the woman-fish is found in G (305-334), although
this is contrary to the source, DC. Later, following DC, G contains
a fearful mediaeval lesson which names those who will perish like
the sailors listening to the siren's song.

> Cil qui aiment tragitaours
> Tumeresses et juglaours,
> Cil ensevent, ce n'est pas fable,
> La procession au deable.   (321-324)

PB (II,172) begins with the important quotation from Isaiah which
he renders in French as: 'La seraine et li diable manront en Ba-
bilone; et li herichons et li haneton[152] mandront en lor maisons
et habiteront." He states that there are three types of sirens; two
are half-woman and half-fish; the third is half-bird. Later he quotes
Phisiologes as saying that the siren is part woman and part bird.
The centaur is also called by PB *sacraire*,[153] and is given a separate
chapter (IV,176) on the war which the horned wild men wage against
him.[154]

Illustrations of the siren, like the sawfish with which it was
sometimes confused, are among the most fascinating of bestiary
drawings because of the ingenuity shown in combining monstrous
features with the figure of a woman. The three types that appear
are women with bird extremities, with fish tail, and finally hybrid
fish and bird features. The text of the oldest illustrated manuscript,
Bern 318, says that the siren has a bird's tail, but the drawing of
a woman standing by a centaur shows her with a coiled, serpent-
like tail.[155] Brussels 10074, f. 146v. shows two bird-tailed sirens
tearing the clothes from a recumbent man while another siren plucks

---

[152] B.N., fr. 834, reads *honocentors*.

[153] And *sagetaire, idem.*

[154] See the description in the Appendix of this study under Wild Man.

[155] The conclusion drawn by Miss Woodruff, *op. cit.*, p. 242, from this
discrepancy does not seem wholly valid in light of the numerous inconsis-
tencies that exist between the text on the siren and its illustration. She says:
"The miniature was not made for the text in which it now stands, but follows the ancient Asiatic story recounted by Ctesias rather than the classical conception which attributes the tail of a bird to the Siren. ... We are

a stringed instrument.[156] Below them is a galloping centaur holding a speared rabbit in his hand.[157] It is odd to see in the Icelandic Bestiary a fish-tailed, bearded woman standing on human legs beside a boat full of lively men. The twelfth and thirteenth century Latin bestiaries from England are scarcely less fantastic in their conception of the siren (Pl. VIII, Figs. 2 and 3). Bodl. 602, f. 10v. depicts two sirens with bird's wings, tails, and taloned feet, playing instruments. A third has a fish's tail and webbed feet. The scene below shows an attack made against two centaurs on whose horse-bodies the dismembered body of a man is tied. In her hand the siren often holds a fish, a mirror, or a comb; more rarely, she grasps the head of a serpent twisted over her shoulder. The onocentaur is usually seen shooting a bow.

SIREN SERPENT.

sirena.

It seems that through some inexplicable reasoning the name of the siren became attached to a winged serpent. Isidore (xii.4.29)

---

in such a case reaching the period when the *Physiologus* text was without religious allegories and consisted merely of accounts of animals, a time before the fourth century. The conclusion seems justified that the *Physiologus* was illustrated as a pure animal book some time between its compilation in the second century and its expansion in the fourth". In the Smyrna *Physiologus* (Strzygowski, *op. cit.*, Pl. II) the siren has an abundant cock's tail. Other than those mentioned in the above discussion, the only drawings of the siren as wholly a woman-bird that I have seen appear in the following manuscripts: Bodl., Douce 167, f. 3v.; Sion College L $\frac{40.2}{L\ 28}$, f. 81v.; Douai 711, f. 32; B.N., lat. 3630, f. 89 (where a long beak disfigures her face), Perrins Bestiary, f. 78.

[156] This scene, not common in Latin manuscripts in the form displayed here —the sirens usually attack the men in their boat— is probably influenced in its conception not only by the text but also by Pliny's brief notice on sirens (x.49.70). He here expresses doubts concerning the reality of sirens, which he classifies as fabulous birds although it has been reported that they exist in India, where they tear men to pieces after charming them with songs. The two legends on the Brussels manuscript read: "Ubi syrene musica sonant ad decipiendos homines" and "Ubi dilaniant eos iam mortuos".

[157] Druce, "Some Abnormal and Composite Human Forms", *op. cit.*, p. 181, suggests that the hare is used here as a symbol of uncleanliness (cf. Pliny viii.55.81), although the portrayal of a centaur holding a hare has elsewhere been connected with maps of the stars.

reports that in Arabia there are serpents with wings which run faster than horses; they also fly. Their poison is such that death follows its bite before pain begins (H iii.47).

A winged serpent with two legs is drawn on Copenhagen, Gl. Kgl. 1633 40, f. 56.

SNAKE.

serpens; *serpens*.

Y (13) quotes Matthew 10:16, "Estote prudentes sicut serpentes, et 'mites' sicut columbe", and says that the snake has three natures.[158] The first nature is that when the snake grows old, its eyes become dim. To renew itself it fasts for forty days, causing its skin to become loose; then seeking a narrow crack in a stone, it pushes through and leaves the old skin behind. The second nature is that on coming to drink in a river, the snake leaves its venom in a hole. The third nature is one applied often to the viper — that of fleeing a naked man and leaping upon a clothed one.

The chapter on the Snake does not belong to the B, B-Is group of manuscripts, though it is found in TH and DC (11), where the three traits are included in the notice on the Viper. It reappears in the enlarged bestiary as the final reptile treated in the section on Reptiles. The snake pushing through a crack is often the last or among the last two or three illustrations in a Second Family manuscript. H (iii.53) keeps the traditional three attributes, and in part following Isidore (xii.4.43) adds that the snake rids itself of blindness by eating fennel. If the snake should taste the spittle of a fasting man, it would die. Pliny, who in reality does not mention it, is quoted as saying that if a snake should escape with its head and only two fingers' length of body, it would still live.

The assumption that the snake becomes young when its skin is shed rests on a misunderstanding of the figurative Greek expression γῆρας ἐκδύνειν, "to shed old age", as in Aristotle (viii 600b 23 ff.).[159] It is similar to the account of the lizard's renewing itself. The depositing of venom is attributed in Aelian (ix.66) to

---

[158] In his edition of Y Professor Carmody adds a fourth nature to the snake. When a man attempts to kill a snake, it exposes its whole body but protects its head. This description is taken from C (Cahier IV,68).

[159] Lauchert, *op. cit.*, p. 16.

the viper before it couples with the muraena, and this trait with a slight change could easily have been transferred to the snake, which in turn gave the attribute of fleeing a naked man, an ancient folk belief, to the viper.[160] Virgil in the Third Georgic (1. 422) speaks of the snake's hiding its head when attacked. This notion might have developed from Aristotle's description of the snake (De part. anim. xi 691b 32 f.). In Pliny (xx.23.95) mention is made of the serpent's use of fennel for skin and sight.

The snake does not appear in French bestiaries before G (501-602), where its DC parentage is evident. G lists three kinds of snakes which are evil and poisonous: the vuivre, described as a viper; the colouvre, to which he attributes the sloughing off of old skin and renewed eyesight; and the dragon, which vomits its poison before drinking and flees the naked man. In PB (II, 217) the fasting and renewal of skin is assigned to a serpent called tiris,[161] from which treacle, an antidote against poison, is made.

Bern 318 pictures a snake squeezing through a wall (f. 11v.), two coiled snakes drinking from a pool (f. 12), and a man piercing a snake with a spear (f. 12v.). The usual scene for the snake is the shedding of the old skin, however little the snake might resemble its own kind (Pl. VIII, Fig. 4). In Epinal 58 (209), f. 72v. (DC) under the title De dracone the third nature of the snake is portrayed —a naked man and a fearsome dragon confront one another.

SPIDER.

aranea; araingne.

The spider is an aerial worm so named because it takes its nourishment *ab aeris*, "from the air". Drawing from its small body a long thread, it never ceases to work upon the stretching of its net (H iii.54; Isidore xii.5.2). TH (175-184) contains verses on the spider's catching flies and on the fragility of its web, easily destroyed by wind. PB (II,212) contains a chapter entitled "L'araingne et la mosche". Here it is said that the saliva of a fasting man kills snakes and spiders should they taste it. After the spider has spun its web,

---

[160] PW, p. 1083.

[161] This serpent is called *tygris* in the *Image du monde de maître Gossouin* (p. 119) and *tiris* in Alexander Neckam (*De naturis rerum* ii.108).

it hides in a corner waiting for a fly or worm. When the fly is caught, the spider runs to devour it and drink its blood.

The spider in the act of spinning its web is illustrated in Arsenal 3516, f. 204v.

STAG.

cervus; cerf.

Both B (29) and Y (43) quote Psalm 41:2 (Vulgate) at the beginning of their chapter: "Sicut cervus desiderat ad fontes aquarum, ita desiderat anima mea ad te, deus." The stag is the enemy of the snake. When the stag knows where the snake is, it fills its mouth with water and spurts it into the hole, drawing the snake out by its breath. The stag then tramples it to death underfoot.

The stag symbolizes Christ, who destroyed the devil-dragon that was unable to endure the fountain of wisdom. G states that the stag signifies a repentant man, who must keep the devil away by vigils and fasting.

In the bestiary attributed to Theobaldus (145-170) the stag is said to have two natures. With its nostrils it draws the snake from its hiding place, swallows it, and burning with poison goes to water to purge itself.[162] There the stag drinks much water, and the poison is overcome. The stag sheds its horns and is renewed. If stags swim across a river to seek food, one rests its chin upon the buttocks of the other, and if the stag in front becomes tired, it goes to the one behind and is carried in turn.[163]

---

[162] DC (13) says that the stag seeks pure water because it is swollen from the poison.

[163] In his marginal summary of the Old English translation of the *Physiologus* attributed to Theobaldus, Richard Morris (ed.), *An Old English Miscellany* (Early English Text Society; London, 1872) records for

> Oc leiged his skinbon
> On odres lendbon.   (359-360)

"Each lays his shin-bone on the other's loin-bone". The Latin in question for this passage reads:

> Portant suspensum gradientes ordine mentum.
> Alter in alterius clunibus impositus.

It is thus apparent that where a "chin-bone" should go a "shin-bone" has been placed.

More is added by H (ii.14) and B-Is (30), based largely on Isidore (xii.1.18,19). *Cervi,* "stags", are so called from their κέρατα, "horns", and they are said to live nine hundred years. When suffering from sickness or old age, they draw out the snake from its hole with their breath and are made well by their meal of poison. Stags can shake off implanted arrows by eating dittany. They marvel at the whistling of a reed pipe and hear well with their ears erect, but not when they are lowered. After telling of their method of swimming so that none grows tired, H alone says that a drink for heart trouble is made from their tears and from bones found in their heart. He then describes the two kinds of stags: the first similar to that of TH, and the second which, after finding and killing a snake, seeks a mountain where it might get food. DC also contains this trait.

The antipathy of the stag and the snake was often mentioned in antiquity. Oppian (*Cyn.* ii.233 ff.) says: "All the race of snakes and deer wage always bitter feud with one another..." (Loeb translation). As is evident above, there are two ways in which the stag extracts the serpent: by spewing water and then drawing forth the serpent with its breath, and by the intake of air alone. Once poisoned, the reason for their seeking water might be explained by the enmity of the crab and the snake, the result of which proves to be an antidote for the stag.[164] Pliny (xxxii.5.19) quotes Thrasyllus as saying that "there is nothing so antagonistic to serpents as crabs; that swine, when stung by a serpent cure themselves by eating them" (Bohn translation). Thus it is fitting for Oppian (*Cyn.* ii.284 f.) to tell of the remedy sought by the stag: "[the stag] seeks everywhere for the dark stream of a river. Therefrom he kills crabs with his jaws and so gets a self-taught remedy for his painful woe" (Loeb translation). Pliny (viii.27.41) mentions the power of the herb dittany, and Aristotle (ix 611b 26) speaks of the stag's being caught by the hunter's pipe playing. Some of H's information is taken from Pliny's chapter on the stag (viii.32.50), and all of the additions in CUL are located there, but probably came by way of Solinus (19.9-19). According to CUL, although the stag is lustful, the female conceives only when the star Arcturus rises. The young are born in a dense wood, kept hidden, and taught to flee over high places. When they hear the barking of hunting dogs, they change their track to the other wind so that their scent will not be carried to the dogs. Their teeth reveal their age, and that they are long lived

---

[164] Wellmann, *op. cit.,* pp. 32-33.

was proved by Alexander the Great, who captured many stags and had them banded. Upon recapture a hundred years later, they were found healthy. Since stags are never feverish, those who eat venison are protected from fever.

The stag is usually shown with its enemy the snake, and often in a double scene where it also appears drinking at a stream (Pl. VIII, Fig. 5 a). The oldest illustration is that of Bern 318, f. 17 where, contrary to the description in many texts of the stag's trampling the snake with its feet, it attacks the snake with its antlers. In addition to a picture of standing stags, Camb., Fitzwilliam Museum 254 (Pl. VIII, Fig. 5 b) depicts the swimming stags supported by their chins. Precise details are included in two manuscripts of DC: in B.M., Sloane 278, f. 51v. a woefully swollen stag drinks at a stream, and in Munich, lat. 6908, f. 81v. while a stag's tongue dips in water, its antlers fall off!

STORK.

ciconia.

According to the *Aviarium* (i.42), repeating Isidore (xii.7.16,17), storks are named after the rattling sound which they make by striking their beaks. They are harbingers of spring, companions of society, and enemies of snakes. When they cross the sea to Asia, crows lead them. They care for their young diligently, and the young reciprocate when their parents grow old.

Ovid noted the stork's clattering noise (*Met.* vi.97), and Pliny (x.23.31,32) speaks of the attention of the young to their parents, a trait also mentioned by Aristotle (ix 615b 23).

In Latin bestiaries the stork is usually portrayed in a recognizable manner with a frog in its mouth (Bodl. 764, f. 64) or with a snake (Bodl. 602, f. 62v.).

SWALLOW.

hirundo; aronde.

The chapter on the swallow does not exist in the Latin B or B-Is versions nor in French bestiaries before PB. Y (42) begins by quoting two Biblical verses in which the swallow is mentioned:

"Sicut 'hyrundo, ita' clamabo, et sicut columba, sic meditabor" (Is. 38:14); "Turtur et hyrundo et cyconia custodierunt tempus 'introitus' sui" (Jer. 8:7). According to Physiologus the swallow breeds once and no more.[165]

The *Aviarium* (i.41) is more detailed concerning this bird. Naming Isidore as its source (xii.7.70), it says that the *hirundo*, "swallow", is so named because it takes its food not sitting, but while *haerendo*, "remaining", in the air. Noisy, flying in circles, it is skillful in building nests and bringing up its young. Endowed with a certain foreknowledge, it deserts buildings that are about to collapse. It is not sought by other birds. The swallow spends the winter overseas, and its return announces the beginning of spring. CUL describes in more detail the construction of the swallow's nest and mentions the medical skill of the swallow in restoring vision to its young.

Aristotle (v 544a 26) does not here agree with the *Physiologus*, because he says that the swallow breeds twice a year. It is unusual for the *Aviarium* not to contain more than a little of what Pliny has said on a certain subject, but concerning the swallow (x.24.34) there is almost no correspondence between the two except the statement that the bird departs for the winter months. The swallow as a harbinger of spring is a belief going back to classical times. The manner in which the swallow obtains mud for its nest is noted in Aristotle (ix 612b 21 f.), while Pliny (viii.27.41) tells how the herb celandine is used in restoring sight.

PB (II,145) speaks of the swallow's gathering food while flying, of its confidence that no bird of prey will catch it because of its swiftness, and of the fact that the bird daubs its nest with earth. If the eyes of the young are torn out, the mother is able to make them see again "mais nus ne set comment ele le fait, ne par coi".

Occasionally the distinctive silhouette of the swallow has been caught by the mediaeval artist in contrast with the usual stereotyped portrayal of birds. One of the best of all bird illustrations is that found in the Hofer Bestiary (f. 5v.) where the two birds gracefully fly in opposite directions.

---

[165] The second Greek version (33 *bis*) printed by Sbordone is the same as Y, but not the first which says that after winter is over the swallow appears in the spring and its song arouses sleepers to work.

SWAN.

olor, cygnus; cygne.

The swan is a bird called by the Greeks κύκνος; however, it is called *olor* in Latin because ὅλος means "all" and the swan's plumes are entirely white, though its flesh is black. It is called *cygnus* from its *canendo*, "singing", because it pours out the sweetness of song with a harmonious voice. Its song is more pleasant because of the swan's long curving neck. In the Hyperborean regions it is said that swans fly to the playing of zithers and that they sing together. Just before its death, the swan sings most sweetly (*Aviarium* i.53; Isidore xii.7.18). CUL quotes from Isidore two lines by the poet Aemilius Macer, who says that the sight of this bird is considered auspicious by sailors.

The only French bestiary under consideration to include the swan is PB (III,233), which recounts the swan's delight in singing to the accompaniment of a harp and the particular excellence of its song during the year in which it is to die.[166]

The voice of the swan presaging its death is an ancient literary allusion; Plato uses this figure in describing the death of Socrates (*Phaedo* 85B). Pliny expresses disbelief in the swan's uttering a mournful song at its death since he tested this story himself (x.23.32). That the swan is a good omen for sailors is found in Servius' commentary on the *Aeneid* (i.393).

In Latin bestiaries the swan is shown in the water, occasionally with a fish in its mouth (Bodl. 764, f. 65v.). In the illustrated manuscripts of PB's long version the swan sings before a harpist.

TIGER.

tigris; tigre.

The account of the tiger in H (iii.1) is a combination of two sources: the etymologies are from Isidore (xii.2.7) and the particular version of capturing the whelps is almost identical with Ambrose's *Hexaemeron* (vi.4.21).

---

[166] For an amusing example of the strange figure which resulted from Brunetto Latini's misinterpretation in *Li Livres dou Trésor* (i.161) of the cause of the swan's death, see the "The Dying Swan - a Misunderstanding" by the present writer in *Modern Language Notes*, LXXIV (1959), 289-92.

The tiger is so called because of its swift flight, for thus the Medes and the Persians call an arrow; and the Tigris River derives its name from the fact that it is the most rapid of rivers. Tigers are spotted and are remarkable for their strength and speed. They come from Hyrcania. The tiger, finding its den empty of whelps, starts in pursuit of the swiftly riding robber. When the thief sees that the tiger's speed will overtake him, he rids himself of the animal by a trick. He throws down a glass sphere, and when the tiger looks at this mirror, it believes its own image is that of the stolen cub. It returns to the empty mirror and settles down as if to nurse the cub. Thus by zeal it loses both revenge and offspring.

Varro had said in the *De lingua latina* (v.100) that both an arrow and the river were called *tigris* in Armenian; it is Solinus (37.5) who attributes the word to the Medes, and Isidore adds the Persians. Pliny's version (viii.18.25) of how to deceive the pursuing tiger is less picturesque than the bestiary account. Instead of mirrors, Pliny has the hunter throw down one cub at a time. The mother returns each one to the lair while the hunter with his prey eventually reaches a ship and safety.

PB (II,140) begins by calling the tiger a kind of serpent.[167] The tiger's delay here is caused by the pleasure it takes in admiring the fair form it sees in the mirror!

The tiger is shown in Latin bestiaries either running after a fleeing horse whose rider holds a stolen cub in his arm, or already deterred by looking into a mirror (Pl. IX, Fig. 1 b). An unusual change takes place in two Third Family manuscripts. In Westminster 22, f. 25 and Bodl. E Mus. 136, f. 18v. the horse or dog-like tiger is looking at a circle hanging from a tree (in the Westminster manuscript a person on foot carries off the cub and in the Bodleian manuscript even the person has disappeared). In Arsenal 3516, f. 200 of PB the tiger, which also gazes at round mirrors attached to a tree, has wings (Pl. IX, Fig. 1 a).

---

[167] This odd statement is probably the result of a confusing reminiscence in copying the initial sentence of two very different chapters in PB: this one on the tiger, which begins, "Une beste est qui est apelée tigre, c'est une manière de serpent", and that on the snake (II,217), described as "Une beste qui est apelé tyris; et c'est 1 serpens dont on fait le triacle".

TURTLEDOVE.

turtur; *turtre, tortorele.*

Y (41) and B (28) begin with a Biblical quotation: "Vox turturis audita est in terra nostra" (Song of Sol. 2:12). Y says that the turtledove will always sit in the desert, and B declares that it is ever faithful to its mate and will not take another should the first one die.

The allegory based on B states that the turtledove is like the holy church, which remained faithful to Christ after his death. GC, referring to the papal interdict then weighing on England, says that many believe that her spouse has abandoned the church.

According to the *Aviarium* (i.25) the bird sometimes descends to the gardens of the poor and to the laborer's fields to collect seeds. It chooses soft and delightful places for its nest. CUL adds from Isidore (xii.7.60) that the bird's name comes from its voice; it loves solitude; and in winter when moulted it lives in hollow trees. To protect its young from attacks by wolves, the turtledove spreads squill leaves over the nest.

The account of the faithfulness and chastity of the turtledove is carried over into the French versions with the addition, first appearing in PT (2547-2555), that the bird, after losing its mate, will not sit on anything green.[168]

Aristotle (viii 600a 20) refers to the bird's hiding in a tree in winter when its feathers drop, and to the bird's single mate (ix 613a 14).[169] From Ambrose's *Hexaemeron* (vi.4.29) comes the reference to the use of squill leaves.[170]

For the most part the illustrations of the turtledove are unsatisfactory, and were it not for the text or title the bird's identity could not be guessed. The Latin bestiaries usually show a conventional standing bird, but in Camb., Corpus Christi Coll. 53,

---

[168] For a short discussion of this characteristic of the turtledove and its relationship with the Spanish romance *Fontefria*, see Leo Spitzer, "Auf keinen grünen Zweig kommen", *MLN*, LXIX (1954), 270-73.

[169] For numerous citations of the turtledove's monogamy by the Church Fathers see Sbordone, *Ricerche*, pp. 131 ff.

[170] Pliny (xx.9.39) quotes Pythagoras as saying that squill suspended over a door will ward off evil spells. Perhaps from this idea arose the story of the turtledove's using it.

f. 202v. the turtledove covers its nest with squills in the presence of a wolf. In some French bestiaries what apparently is a barren tree is introduced.

UNICORN.

unicornis, monoceros, rinoceros; *unicorne.*

The unicorn, doubtless the best known of fabulous animals, has had a varied career and, according to one of the authorities on its history, the spread of popular tradition concerning this beast as distinguished from the learned was by means of the bestiaries.[171] A significant place in literature and art is occupied by this animal of hybrid ancestry — descending from the Indian ass among the Greeks and the horse among the Greeks and Romans, and from the goat in the *Physiologus,* with the rhinoceros strain constantly reappearing.

Y (35) begins with a Biblical quotation: "Moyses de monoceraton in Deuteronomio dixit, benedicens Ioseph: 'Primitivus' tauri 'species' eius, cornua unicornui cornua eius" (Deut. 33:17).[172] The unicorn is described as a small animal like a kid, very fierce, and having one horn in the middle of its head. No hunter can approach this animal because of its strength, so in order to catch it a virgin is placed in its path. When the unicorn springs into her bosom, she warms and suckles it.[173] It is then caught and taken to the king's palace. H (ii.6) makes no major change in the account. The virgin, on perceiving the unicorn, uncovers her bosom, the sight of which causes the animal to lose its wildness. It then puts its head in her lap and goes to sleep. B-Is (16) omits any reference to the bosom, and says merely: "At ille visa virgine complectitur eam et dormiens in gremio eius..." It differs also from H by including a passage from Isidore, which will be discussed later.

Allegorically, except for PT's markedly different minor details, all versions agree that Christ is the spiritual unicorn who, descending into the Virgin's womb, was incarnate, was captured by the Jews, and condemned to death. The unicorn's horn symbolizes Christ's unity with his father; the animal's fierceness, the inability of heaven-

---

[171] Odell Shepard, *The Lore of the Unicorn* (London, 1930), p. 47.

[172] In B (16) the Greek name *monosceros* and the Latin *unicornis* are first given.

[173] This last is not contained in B.

ly powers to know Christ and of Hell to hold him; its small size signifies Christ's humility in assuming humanity; and its kid-like appearance represents Christ's being made in the likeness of carnal sin. PT, before straying far from the traditional allegory, explains that the virgin's bosom represents the church and that the kiss signifies peace.

The earliest account of the unicorn is in the *Indica* of Ctesias, a fifth century B.C. Greek who was physician at the court of Darius II, king of Persia.[174] The one-horned animal is there called a wild ass from India, and seems in part to be based on the Indian rhinoceros. Pliny (viii.20.29) in describing the rhinoceros mentions that it is a natural-born enemy of the elephant and that when fighting, it aims at its adversary's belly. Then after speaking of other animals which have one horn, he mentions (viii.21.31), without realizing that he is again referring to the rhinoceros, the fierce *monoceros* which has the head of a stag, elephant's feet, the tail of a boar, and a horse's body. It makes a deep lowing sound, and has a black horn two cubits in length in the middle of its forehead. Solinus (52. 39,40) later repeats this description. Much importance is given to the unicorn in Aelian (iii.41; iv.52; xvi.20), who in the third passage which treats the *cartazon* seems to be describing a rhinoceros with an antelope's horn.[175] Its excessive fierceness is stressed, and the fact that "the young are sometimes taken to the king to be exhibited in contests on days of festival because of their strength". Elsewhere (xvii.44), in his chapter on the rhinoceros's combat with the elephant, Aelian refuses to describe the rhinoceros because it is familiar to the Greeks and Romans. Thus the two animals, the rhinoceros and the unicorn, which were originally one, are separated.

There are seven references, all in the Old Testament, to the unicorn in the King James Version of the Bible,[176] the word being derived from the Septuagint translation μονόκερως of the Hebrew *Re'em*. The beast is always alluded to as being strong and fierce. Where then did the story of the small goat-like animal captured by a virgin originate?

Sbordone believes that the small unicorn was invented by the Christian compiler in order to facilitate the capture by the virgin, and is thus another example of many cited by this scholar where

---

[174] Migne, *Patr. Gr.*, CIII, Col. 226.
[175] Shepard, *op. cit.*, p. 36.
[176] Num. 23:22; Deut. 33:17; Job 39:3-12; Ps. 22:21; Ps. 29:6; Ps. 92:10; Is. 34:7. In the Vulgate the unicorn is also found in Ps. 77:69.

the allegory influenced the zoological account.[177] A passage identical with that of the *Physiologus* is found in the Commentary on Basil's *Hexaemeron,* long attributed to Eustathius of Antioch.[178] The next appearance, important for its later influence, is in Isidore (xii.2. 12,13), where the opening words show the confusion existing between the rhinoceros and the unicorn: "Rhinoceron a Graecis vocatus. Latine interpretatur in nare cornu. Idem est monoceron, id est, unicornu..." In this notice the unicorn is spoken of as fighting with the elephant — a trait taken from ancient descriptions of the rhinoceros — and as captured by a virgin. In the expanded Latin bestiaries of the Second Family and thereafter, there exist two chapters on the Unicorn: the first, *De unicorni,* describing the virgin-capture; and some chapters later, *De monocerote,* transcribing Solinus' frightful horse-bodied rhinoceros.

It appears that the ultimate origin of the role played by the virgin in ensnaring the unicorn is lost, although non-Christian tradition has probably influenced the story.[179] In the Syriac version the emphasis is upon sexual attraction by the girl: "...Then the girl offers him her breasts, and the animal begins to suck the breasts of the maiden and to conduct himself familiarly with her..."[180] Definite traces of using physical means to draw the unicorn to his ultimate death can be found in the oldest Old French version.[181]

The *Physiologus* tradition is at once apparent in PT (393-418) where the unicorn is described as resembling a *buket,* "little goat". The physical attraction of the virgin is also stressed, for when the hunters find the unicorn's repair,

> La met une pulcele
> Hors del sein sa mamele;
> E par l'odurement
> Monosceros la sent;
> Dunc vient a la pulcele
> Si baise sa mamele,
> En sun devant se dort,   (403-409)

---

[177] Sbordone, *Ricerche,* p. 59.

[178] Migne, *Patr. Gr.,* XVIII, Col. 744.

[179] Shepard, *op. cit.,* p. 66.

[180] *Ibid.,* p. 49. Although Shepard states that this citation is from the Syriac version, it is in reality from the Arabic. See J. P. N. Land, "Scholia in Physiologum Leidensem", *Anecdota Syriaca,* IV (1875), 147.

[181] In Richard de Fournival's *Bestiaire d'Amour* (42,7) and the bestiaries related to this the odor of the virgin is stressed as the power which attracts the unicorn.

GC (1375-1416), based on B-Is, includes Isidore's contributions, but instead of the horn being the sharp instrument which pierces the elephant's belly, it is the unicorn's foot with a sharp nail which cuts the elephant.[182] On seeing the virgin, the unicorn crouches on her lap, then plays so much with her that it falls asleep. When captured, it is driven before the king. In G (239-264) a beautiful virgin is sent as a decoy into the desert where the beast is accustomed to live.[183] PB's account (II,220) seems to be based on Solinus' rhinoceros-unicorn description, except that here the animal has a high and clear voice instead of Solinus' "mugitu horrido".

The many illustrations of the unicorn show the same instability as to physical characteristics that has been noted in the texts. The oldest Latin *Physiologus* picture in Bern 318 (Pl. IX, Fig. 2 a) portrays the animal as a goat with a curving horn, its muzzle placed in the hands of a standing woman. This pose has not been seen elsewhere.[184] Brussels 10074, f. 147 contains two scenes: in one a woman leads a large, clovenhoofed unicorn toward a king seated within a stylized palace; in the other the woman is seated on a folding chair with the standing animal's head in her lap. Twelfth and thirteenth century manuscripts apparently all show the hunter

---

[182] An explanation for the transference of the cutting from the horn to the foot can be discovered by comparing Isidore's Latin (xii.2.12) with GC's Old French. Isidore says: "... id est, unicornus, eo quod unum cornu in media fronte habeat pedum quatuor ita acutum et validum ut quidquid impetierit, aut ventilet aut perforet. Nam et cum elephantis saepe certamen habet, et in ventre vulneratum prosternit." GC expresses the passage in this way:

> Tant a le pe dur e trenchant
> E l'ongle del pe si agu,
> Que ren n'en poet estre feru,
> Qu'ele ne perce e qu'el ne fende,
> N'a pas poeir que s'en defende
> Li olifanz, quant le requert:
> Car desoz le ventre le fert,
> Del pe trenchant com alemele
> Si forment, que tot l'esboële   (1384-1392)

It is evident that the *pedum* meaning "foot" in the sense of measurement in Isidore was understood as *pedem,* an anatomical "foot" in GC. This erroneous interpretation has been pointed out by Lauchert, *op. cit.,* p. 146, n. 1.

[183] Aelian (xvi.20) says of the unicorn's habitat: "Desertissimas regiones persequitur, simul et solitarius errat."

[184] The resemblance between this unicorn and those depicted in the Utrecht Psalter is noted in Chapter IV of this study.

attacking the unicorn with all manner of weapons while the animal either kneels or awkwardly crouches with its forefeet in the girl's lap (Pl. IX, Fig. 2 b). The unicorn itself has many forms — from that of a small dog with a straight horn to a large graceful animal with a curved horn. It is pertinent to note the position of the unicorn's head in the Leningrad Bestiary — it is inside the woman's dress. This would seem to be evidence that the word *sinus* was understood as meaning "bosom" rather than "lap" as it is sometimes translated into English from the Latin. Except for his elephant-like feet the *monoceros* usually resembles the unicorn (Pl. IX, Fig. 2 c).

In the French manuscripts, as has been seen in other cases, the compositions are more simple, but the variations in the depiction of the unicorn are as great as in the Latin. In the Copenhagen manuscript of PT (f. 15) the unicorn is a goat-like creature in the lap of a maiden, while in two GC manuscripts (Bodl., Douce 132, f. 70 and Camb., Fitzwilliam Museum, McLean 123, f. 40v.) the woman is unclad. All of these animals are far removed from the spirited, finely proportioned unicorns of later heraldry.

VIPER.

vipera; wivre, woutre.

The account of the Viper underwent bizarre changes in the course of time and, because in the Greek and the early Latin versions the chapter on the Viper was followed by that on the Snake, and since the viper is a type of serpent, some of the latter's characteristics were attached to the former and have so remained.

Y (12), after the expected mention of "Generatio viperarum" (Matt. 3:7), recounts that the male viper has the appearance of a man, and the female of a woman to the waist, but below the navel she has a crocodile's tail. Thus the female has no place to conceive. When the male injects his seed into her mouth, she cuts off his member and he dies. The young vipers have no place through which to be born, so they open their mother's side and come forth, killing her.

This is the extent of the oldest Latin versions, though earlier literature has elements of the story. In the *Theogony* (l. 297 ff.) Hesiod describes the monster daughter of Callirhoé and calls her Echidna (ἔχιδνα being the Greek name for "viper"). The crocodile's tail would seem to show Egyptian influence.[185] The viper's impregnation through the mouth (although the viper in these cases is wholly reptilian) and the subsequent death of all are found in Herodotus (iii.109), Pliny (x.62.82), and Aelian (i.24).

With H (ii.21) and DC (11) — the viper is not included in B — the human appearance is wholly lacking and conception occurs when the mouth of the male is inserted in that of the female. She then cuts off his head.[186] The ensuing traits listed by Hugo and DC rightfully belong to the snake and are discussed in that chapter except for one which was transmitted to PB — the third nature of the viper which is to flee a naked man and to leap upon one who is clothed. An addition which describes the coupling of the viper and the muraena apears in CUL, and finds its origin in Ambrose's *Hexaemeron* (v.7.18).

G's account (507-521) of the conception and birth of the viper is normal. PB has two chapters on the viper, perhaps thinking of it as two different creatures. In the chapter telling of the birth, the animal is called a *wivre* (II,134); in the one connected with its fleeing a naked man, it is called a *woutre* (II, 143).[187]

Of interest is the oldest illustration of the viper, Bern 318 (Pl. IX, Fig. 3), because it portrays the male and the female as humans with tails. The male points with his finger to the mouth of the female. In Morgan 832, f. 5v. of DC the viper is drawn as a snake which swallows the head of another, while below two small snakes emerge from their mother's back. More commonly the viper does not appear as a snake but with wings and two feet (Pl. X, Fig. 1). To illustrate PB's *woutre* a beast runs from a naked man in Arsenal 3516 (Pl. VIII, Fig. 6).

VULTURE.

vultur; voltoir.

The substance of Y's chapter (32) on the Vulture is entirely different from the later Latin and French attributes of this bird. Y says that the vulture is found in high places. When pregnant it goes to India and there finds the eutocium stone (λίθος εὐτόκιος).

---

[185] PW, *op. cit.*, p. 1082.

[186] Cf. Isidore (xii.4.10,11).

[187] In a mediaeval manuscript it would not be very difficult to mistake the appearance of the word *wivre* for *woutre.*

This stone is similar to a nut in size,[188] but when moved it sounds like a bell. At the moment of birth the vulture sits on the stone and delivers without pain.

Legendary attributes of the eagle have here been connected with the vulture.[189] Writers of antiquity called the stone mentioned in the account of the vulture the eagle-stone (λίθος ἀετίτης or *lapis aetites*), and it had two functions. Eagles were said to use it in the construction of their nest in order to anchor it more firmly.[190] Its other use as an aid in birth was more generally referred to.[191] Wellmann has suggested the possibility of the original name λίθος ἀετίτης being altered to λίθος εὐτόκιος.[192]

The *Aviarium* (i.38) mentions none of the above story of the vulture, but describes it as a bird of prey which flies down to seize a dead body and which follows armies to feed on corpses. This habit is mentioned also in Ambrose's *Hexaemeron* (v.23.81). Isidore (xii.7.12) is quoted as saying that the *vultur* is thought to receive its name "a volatu tardo", "from its slow flight". CUL's chapter on the Vulture largely differs from that in the *Aviarium*, though it notes the bird's ability to predict the number of dead in battle. From Ambrose's *Hexaemeron* (v.20.64) is drawn the statement that the vulture's offspring are engendered without mating and live to be a hundred years old.[193]

The vulture is not mentioned in any French bestiary previous to that of PB (II,146), which tells of its desire to follow armies for their carrion that it can smell from a distance of a three day journey.[194] When a corpse is found, the vulture first eats the eyes

---

[188] Y reads "nix similis magnitudine", but this must be a misprint for "nux" since the Greek text reads κάρυον.

[189] See article by A. A. Barb, "Birds and Medical Magic: 1. The Eagle-Stone; 2. The Vulture Epistle", *Journal of the Warburg and Courtauld Institutes*, XIII (1950), 316-22.

[190] Horapollo (ii.49).

[191] The explanation of the connection between this stone and pregnancy is found in Pliny (x.3.4): "This stone has the quality also, in a manner, of being pregnant, for when shaken, another stone is heard to rattle within, just as though it were enclosed in its womb" (Bohn trans.).

[192] Wellmann, *op. cit.*, p. 88.

[193] The manner of engendering is explained in Horapollo (i.11): "...when the vulture hungers after conception, she opens her sexual organs to the North Wind and is covered by him for five days..."

[194] Cf. Pliny (x.6.7). PB repeats this in another chapter on the Vulture (IV,80), where it is found together with the keen sighted worm *lieus*, and though it is not explicitly stated, the vulture here signifies the sense of smell

and then extracts the brain through the orbit (this trait is also linked with the raven). Needless to say, it is considered a dirty bird.[195]

The carniverous tendencies of the vulture are evident in most of its illustrations. In B.M., Harl. 3244, f. 51 a large bird places its talons on the shoulder of a corpse; in Camb., Trinity Coll., R.14.9, f. 100 it has part of a human arm or leg in its beak; and in Bodl. 764, f. 61 two vultures tear the flesh of animals.

In the illustrations of French bestiaries special emphasis is placed on the vulture's flying behind armed knights on horseback as in Arsenal 3516 (Pl. X, Fig. 2).

WEASEL.

mustella; *mustelete, mostoille, belette.*

Although in most versions prior to the enlarged Second Family the account of the Asp is included in the same chapter with the Weasel, each is discussed separately here. Y (34) begins with the admonition that according to the law, weasels are not to be eaten because they are dirty animals (Lev. 11:29). The weasel is said to conceive at the mouth and to give birth through the ear. Y then contains a statement not found in B (26) — if mating takes place through the right ear, a male will be born; if it is through the left ear it will be a female. The weasel signifies those who willingly receive the seed of the divine word, but then hide what they have heard.

H (ii.18) and B-Is (27) differ slightly concerning the weasel's conceiving. Here it is reported that some say it conceives by the ear and gives birth through the mouth, while others say that conception takes place through the mouth and birth through the ear.[196] The rest of H's and B-Is's remarks are taken from Isidore (xii.3.3) although the etymology of the name *mustela* is changed. According to H and B-Is the animal is so called because it is like a *mus longus*, "long mouse", for *telos* (τέλος) means "long" in Greek. Isidore

---

as the *lieus* does sight. The chapter follows that on the Elements and Senses with which, in all probability, it should form a whole.

[195] In the Waldensian Bestiary (15) we return to the ancient story of the eagle-stone.

[196] Cahier (II,149) offers the following suggestion to explain the mouth-ear ambivalence: "Des équivoques comme 'aurem' and 'orem' (pour 'os') auront pu occasioner les différentes versions également merveilleuses qui couraient sur l'enfantement de la mustoile..."

PRINCIPAL SUBJECTS   187

used the word *telum*, "spear", in this place and said that the animal
derives its name from its spear-like length. Weasels are clever because they move their young from place to place after birth. They
pursue snakes and mice. There are two types of weasels: one, which
is called by the Greeks *ictidas* (Ἰϰτις), lives in the woods; the
other wanders about in houses. Not found in Isidore is the statement
that weasels are skillful in medicine, knowing how to revive their
young should they be killed.[197]

The belief that weasels give birth through the mouth was already
being refuted by Aristotle, who attributes this misconception to
Anaxagoras (*De gen. an.* iii 756b 15).[198] A surprising assertion is
found in Pliny (x.65.85), who states that the lizard gives birth at
the mouth, and says that Aristotle denies it. This confusion apparently arose from mistaking the word γαλῆ, "weasel", for γαλεώτης,
"spotted lizard".[199] The impregnation through the mouth and birth
at the ear, instead of the contrary, is a later version which might
have been influenced by tales such as those about the viper.[200]

In the Old French accounts the same hesitancy as to the location
of each action is sometimes found. The short passages in PT (1217-
1228) and G (1137-1144) only mention the conception at the mouth
and birth at the ear. GC (2419-2436) repeats this and questions in-
credulously:

> Sont cil fols, qui vont affermant,
> Que ele receit e espant
> La semence parmi l'oïe?
> Sëurement   ceo n'i a mie.   (2433-2436)

PB (II,148) and GC alike speak of moving the young about, and
the former records the old belief that the weasel can revive its dead
offspring.

---

[197] No ancient references have been found referring to the belief that
the weasel knew the herb of life, but mention is made of this power in
Marie de France's lai *Eliduc* and in Giraldus Cambrensis, *Topographia
Hibernica* (i.27). See Thomas S. Duncan, "The Weasel in Religion, Myth
and Superstition", *Washington University Studies* (Humanistic Series), XII
(1924), 62-65.

[198] That the opinion was current in Egypt also is seen in Plutarch
(*De Is. et Os.* 381A): "There are still many people who believe and declare
that the weasel conceives through its ear and brings forth its young by
way of the mouth, and that this is a parallel of the generation of speech."

[199] Wellmann, *op. cit.,* p. 28.

[200] PW. *op. cit.,* p. 1087.

Many illustrations merely portray a lengthy, long-nosed animal, but others have more individual renditions (Pl. X, Fig. 3). In B.M., Sloane 3544, f. 20v. two weasels stand mouth to mouth as a tiny weasel pops from the ear of one of them. Rats and dragons flee from the weasel in B.M., Royal 12 F. xiii, f. 44; in Camb., Corpus Christi Coll. 22, f. 164 the weasel transports its young and stands over one that must be dead. Finally in Sion College L $\frac{40.2}{L28}$ f. 96v. birth is taking place at the weasel's mouth.

WETHER and RAM.

vervex, aries.

The *vervex*, "wether", is called either from its *vires*, "strength", being stronger than other sheep, or because it is a *vir*, "male", though castrated, or because having *vermes*, "worms", in its head it becomes excited by their itching and strikes other sheep. *Aries*, "ram", comes from "Αρεος, "Ares" — that is, from Mars, by which name male sheep are called; or the name comes from the fact that rams were first sacrificed by the Gentiles on the *ara*, "altar". Isidore (xii.1.10,11) and H (iii.14) cite a line from the fifth century Christian poet Caelius Sedulius: "Aries mactatur ad aram." However, an objection to the derivation of *aries* from *ara* is expressed in H, where after this quotation is written, "Sed id minus placet, et primae syllabae quantitas non approbat".

An illustration in Camb., University Library Kk.4.25 shows a wether wearing a bell, behind a flock of sheep before which two rams butt one another. Above the head of the hooded shepherd are the words: "Ha ha ware le corn."

WOLF.

lupus; lup, leu.

This is among the longer of the chapters not found in the oldest Latin texts, and it is one of the earliest additions to B-Is. The text in H (ii.20), differing from Isidore's etymology (xii.2.23,24), says that the Greeks call the wolf *lycos* (λύκος), and that *lux* (λύκη) means "morning light", a derivation fitting for plunderers. Thus prostitutes are termed wolves because they lay waste their lover's goods. Others say that wolves are called *lupos* because they resemble *leopedes*, since like the lion their strength lies in their feet, and what they trample on does not live. Countrymen say that a man loses his voice if first seen by a wolf, but if it is the animal which is first seen, it loses its fierceness. The wolf is strongest in its shoulders or mouth, weakest in its loins, and it is unable to turn its neck backwards. The wolf lives from prey, from the earth, and sometimes from the wind. During the year the wolf does not couple more than twelve days, the female bearing its young only in May when it thunders. Nourishment for the young is not sought nearby but at a distance, where the wolf approaches the sheep fold like a tame dog, and lest the watch dogs smell its evil breath and awaken the shepherds, it goes against the wind. If a branch should break underfoot, the wolf fiercely bites its foot as punishment. In the night its eyes shine like lamps. Solinus (2.36) is given as H's authority for stating that at the tip of a wolf's tail is a tuft of hair having amatory power which the wolf bites off if capture is imminent. H explains what a man should do when his voice has been silenced by the wolf. He should remove his clothes, trample the ground underfoot, and taking two stones in his hand, should strike them together. CUL includes Isidore's derivation of the Greek name λύκος, "wolf", from λύσσα, "rage".

Pliny (viii.22.34) attributes the wolf's ability to cause muteness to a harmful power in its eye, mentioning also the tuft of hair in its tail and its coupling only twelve days, although Aristotle (vi 580a 15) states that this length of time applies to the lying-in period of the she-wolf.

The short version of PB's bestiary, as has been pointed out in Chapter III, is based on B-Is with the addition of notices on the Dog and on the Wolf. PB's text on the Wolf (IV,71) is a translation of the chapter found in B.M., Stowe 1067, and resembles H down to the reference to Solinus.

In the Latin bestiaries the scene most often represented is that of the wolf approaching the sheep fold and biting its noisy paw (Pl. X, Fig. 4), although a wolf fleeing with a lamb in its jaws appears on B.M., Harl. 3244, f. 44. Apparently a unique illustration is that found in B.M., Royal 12 F. xiii, f. 29, where the man who has been rendered speechless stands upon his shirt and holds a large stone in each hand in an effort to recover his voice.

WOODPECKER.

picus; espech, espec.

In the same chapter with the Magpie (*pica*), H includes the Woodpecker (iii.32), saying that *picus* takes its name from Saturn's son Picus who used this bird in auguries. The woodpecker is considered divine since no spike can remain inserted in any hole where the woodpecker has made its nest. Isidore (xii.7.47) adds that this is the *picus Martius*.[201]

Why should these accounts omit the essential factor in this bird's ability to reopen its plugged hole — the existence of an herb which the woodpecker applies to the peg and which Pliny mentions (x.18.20)?[202] The woodpecker and the herb which unplugs whatever it touches is included in PB (II,160).

Unillustrated in the Latin bestiaries because of its inclusion in the section on the Magpie, in Arsenal 3516 the woodpecker flies with a leaf in its beak to a hole that is closed with a spike (Pl. X, Fig. 6).

YALE.

eale; centicore.

Not found in Isidore but in Solinus (52.35) and in H (iii.10), the yale had an even more composite career in the vernacular than in its original form as an animal large as a horse, black in color, with an elephant's tail and the jaws of a wild boar. Its horns are unusually long and adapted for movement, for one is folded back while the other fights. If a blow damages the sharp point of one, the other is used.

In PB (III,223), the only French bestiary to include this animal, the yale is called a *centicore* and is said to dwell in the deserts of

---

[201] For a discussion of the woodpecker which was equated with Zeus, and its relation with Picus, son of the king of Latium, who was metamorphosed into a woodpecker (Ovid, *Met.* xiv.390; Virgil, *Aen.* vii.197), see W. R. Halliday, "Picus-who-is-also-Zeus", *Classical Review*, XXXVI (1922), 110-12.

[202] In the *Koiraniden* the same story is told of the δενδροκολάπτης, Wellmann, *op. cit.*, p. 97.

India. After speaking of its mobile horns, which are longer than four arm lengths, it is described as having a muzzle like the bottom of a barrel, thighs and chest like a lion, the feet and body of a horse, an elephant's tail, and a voice like a man's. The basilisk (*basilecoc*) hates the *centicore*, and finding it asleep, pricks it between the eyes so that it swells until its eyes burst from its head and the animal dies from poison.

George Druce has pointed out that this French version is composed of details from the yale and the leucrota as can be seen by comparing Pliny's consecutive remarks on each (viii.21.30).[203] In the *Image du Monde* of Gossouin de Metz the leucrota is called a *centicore* and its chapter is followed immediately by that of an unnamed animal whose description is that of the yale; and in some manuscripts, by the omission of line, the two accounts are fused into one under the title *centicore*.[204] The derivation of the word *centicore*, though perhaps having an origin similar to *manticora*, is uncertain,[205] and the connection of the basilisk with this story is unaccountable.

A humorous variation appears in some of the Latin texts and illustrations. Some read *maxillis aprinis*, "with jaws of a wild boar", and others *maxillis caprinis*, "with jaws of a goat". This difference is perhaps attributable to the vagaries of a scribe who thought it inconsistent for a beast to have both horns and tusks and therefore added a *c* to the name of the animal.[206] In illustrations where the reading is the former, the yale has no beard, but with the latter it

---

[203] George C. Druce, "Notes on the History of the Heraldic Jall or Yale", *Archaeological Journal*, LXVIII (1911), 173-99.

[204] *Ibid.*, pp. 184-85.

[205] *Ibid.*, p. 185, n. 2. A possible solution of the formation of the word *centicore* might lie in another work by Pierre de Beauvais. In the résumé of Pierre's *Mappemonde* given by Langlois, *La Connaissance de la nature* (1927), p. 185, four animals are described by Pierre in the same order in which they appear in Pliny (viii.21.30), Solinus (52.34-37), and Honorius Augustodunensis (PB's probable source. See Migne, *Patr. Lat.*, CLXXII, Col. 124). Their Latin names in succession are: *leucrocota* (Honorius writes *ceucocroca*), *eale*, *indicus taurus*, and *mantichora*, but Pierre calls them *santicora* (with a variant given by Langlois, *ceucrocata*. This reading is found in B.N., nouv. acq. fr. 13521, f. 63 b), *eale, taureaux, manticore*. It is not improbable that the first syllable *ceu-* was thought to be *cen-* in a manuscript, and then because they are almost homonymous, it was written *san-*. And doubtless the ending *-ticora* was the result of a scribe's attention dropping down a few lines in his source to the word *manticora*.

[206] *Ibid.*, p. 178.

usually has a goatee. In Morgan 81 (Pl. X, Fig. 5) the rather graceful animal has two horns, one pointing up and the other forward, while in Camb., Fitzwilliam Museum 254, f. 20 the yale resembles a bearded bull with thick horns pointing ahead and behind. In Arsenal 3516, f. 206v. of PB the basilisk, with cock's body and snake's tail, pricks with its beak the seated *centicore*, but in the Montpellier manuscript of PB, a normal, horizontally horned yale is drawn with no basilisk in view.

ILLUSTRATIONS

Since iconographical considerations were of more importance for this study than stylistic, certain liberties have been taken with the tracings made of the manuscript illustrations: landscapes were eliminated where they did not affect the scene; frames were omitted; depiction of clothing was simplified; the size of the original was reduced or enlarged; and in the illustration of the Beaver two separate scenes were juxtaposed. In no case, however, have any of the essential features been changed as far as these were visible on microfilms or photographic reproductions. In spite of the incomparable aid given by the various processes of reproduction in existence today, it is still true that nothing can wholly take the place of seeing the manuscript itself — with its dark letters and rubrics, its lively drawings in outline, or its scenes in many colors.

The animals in the following illustrations represent a personal choice based, among other reasons, on scenes that are most typical of the familiar or of the lesser known of the *Physiologus* and bestiary creatures. It will be noted that the majority are taken from Latin rather than French manuscripts. This can be explained by recalling that the latter in many cases contain small, rather unimaginative pictures usually dating from the thirteenth century; they include a number of interesting details but their general excellence is not equal to that found in the wide variety of extant Latin manuscripts.

SOURCES OF ILLUSTRATIONS

Plate I.

    1. AMPHISBAENA
        a. Bodl. 764, f. 97.
        b. Cambridge, Fitzwilliam Museum 254, f. 41v.
    2. ANT. B.N., fr. 14969, f. 17. (GC).
    3. ANTELOPE. Brussels, Bibl. Roy. 10074, f. 141.
    4. APE. Bodl. 602, f. 18v.
    5. ASP. B.M., Harl. 3244, f. 61v.

Plate II.

    1. ASPIDOCHELONE
        a. Bodl. 602, f. 22v.
        b. Oxford, Merton Coll. 249, f. 8. (PT).
    2. BASILISK. Cambridge, Corpus Christi Coll. 53, f. 205.
    3. BARNACLE GOOSE. Paris, Arsenal 3516, f. 205. (PB).
    4. CALADRIUS. B.M., Egerton 613, f. 34. (GC).
    5. BEAVER. B.M., Royal 2 B. vii, f. 101v. and 102.

Plate III.

    1. CROCODILE. Morgan 81, f. 70.
    2. CRANE. Bodl. 764, f. 62.
    3. CALADRIUS. Morgan 832, f. 10.
    4. DRAGON
        a. B.N., lat. 3630, f. 93.
        b. B.M., Harl. 3244, f. 59.
    5. EAGLE. Cambridge, Corpus Christi Coll. 53, 198v.

Plate IV.

    1. ELEPHANT
        a. Bodl., Laud Misc. 247, f. 163v.
        b. B.M., Harl. 4751, f. 8.
    2. FOX. Bodl. 602, f. 12v.
    3. GRIFFIN. Bodl. 764, f. 11v.
    4. MANDRAKE. B.N., fr. 14969, f. 61v. (GC).

Plate V.

1. HEDGEHOG. B.M., Royal 12 F. xiii, f. 45.
2. HYDRUS AND CROCODILE
   a. Copenhagen, Royal Lib. 3466, f. 21. (PT).
   b. Cambridge, Corpus Christi Coll. 53, f. 206.
3. LIZARD. Munich, lat. 6908, f. 81v.
4. IBIS. Bodl. 602, f. 11v.
5. HYENA. B.M., Harl. 4751, f. 10.

Plate VI.

1. LION
   a. Copenhagen, Royal Lib., 3466, f. 10. (PT).
   b. Cambridge, Corpus Christi Coll. 53, f. 189.
2. MANTICORA. Bodl. 764, f. 25.
3. OSTRICH. Canterbury Lit. D 10, f. 131.
4. PANTHER. Bodl. 602, f. 20.

Plate VII.

1. PHOENIX. Bodl., Douce 88 A, f. 20.
2. ONAGER. Munich, lat. 6908, f. 80.
3. SALAMANDER. B.M., Royal 12 C. xix, f. 68v.
4. PELICAN. Bodl. 764, f. 72v.
5. PERIDEXION TREE. Bodl. 602, f. 29.
6. SAWFISH. Morgan 81, f. 69.

Plate VIII.

1. SAWFISH
   a. Copenhagen, Royal Lib. 3466, f. 43v. (PT).
   b. Brussels, Bibl. Roy. 10074, f. 142.
2. SIREN AND ONOCENTAUR. Bodl., Douce 167, f. 3v.
3. SIREN. Cambridge, Sidney Sussex Coll. 100, f. 38v.
4. SNAKE. Bodl. 764, f. 100v.
5. STAG
   a. Bodl. 764, f. 17v.
   b. Cambridge, Fitzwilliam Museum 254, f. 10.
6. VIPER. Paris, Arsenal 3516, f. 200v. (PB).

Plate IX.

    1. TIGER
        a. Paris, Arsenal 3516, f. 200. (PB).
        b. B.M., Add. 11283, f. 1.
    2. UNICORN
        a. Bern, Burgerbibliothek 3,18, f. 16v.
        b. B.M., Royal 12 F. xiii, f. 10v.
        c. B.M., Harl. 3244, f. 42v.
    3. VIPER. Bern, Burgerbibliothek 318, f. 11.

Plate X.

    1. VIPER. B.M., Sloane 278, f. 51.
    2. VULTURE. Paris, Arsenal 3516, f. 200v. (PB).
    3. WEASEL. B.M., Royal 2 B. vii, f. 112v.
    4. WOLF. B.M., Sloane 3544, f. 13.
    5. YALE. Morgan 81, f. 39.
    6. WOODPECKER. Paris, Arsenal 3516, f. 201v. (PB).

PLATE I

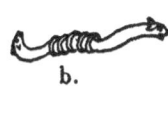

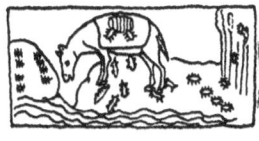

2. Ant

1. Amphisbæna

3. Antelope

4. Ape

5. Asp

PLATE II

a.

2. Basilisk

1. Aspidochelone

b.

3. Barnacle Goose

4. Caladrius

5. Beaver

PLATE III

1. Crocodile

2. Crane

3. Caladrius

a.

b.

4. Dragon

5. Eagle

PLATE IV

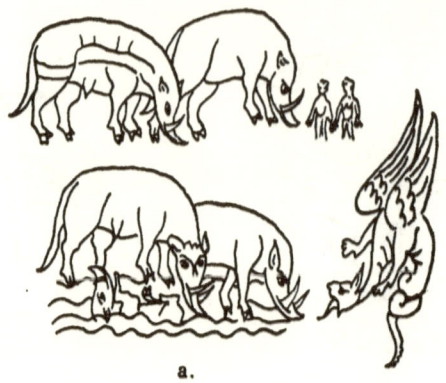

a.

1. Elephant

b.

2. Fox

Mandrake

3. Griffin

PLATE V

1. Hedgehog

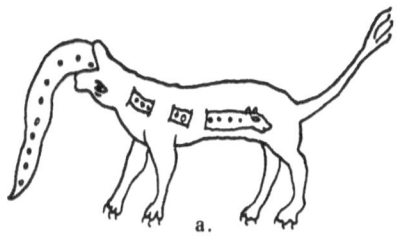

a.

2. Hydrus & Crocodile

b.

3. Lizard

4. Ibis

5. Hyena

PLATE VI

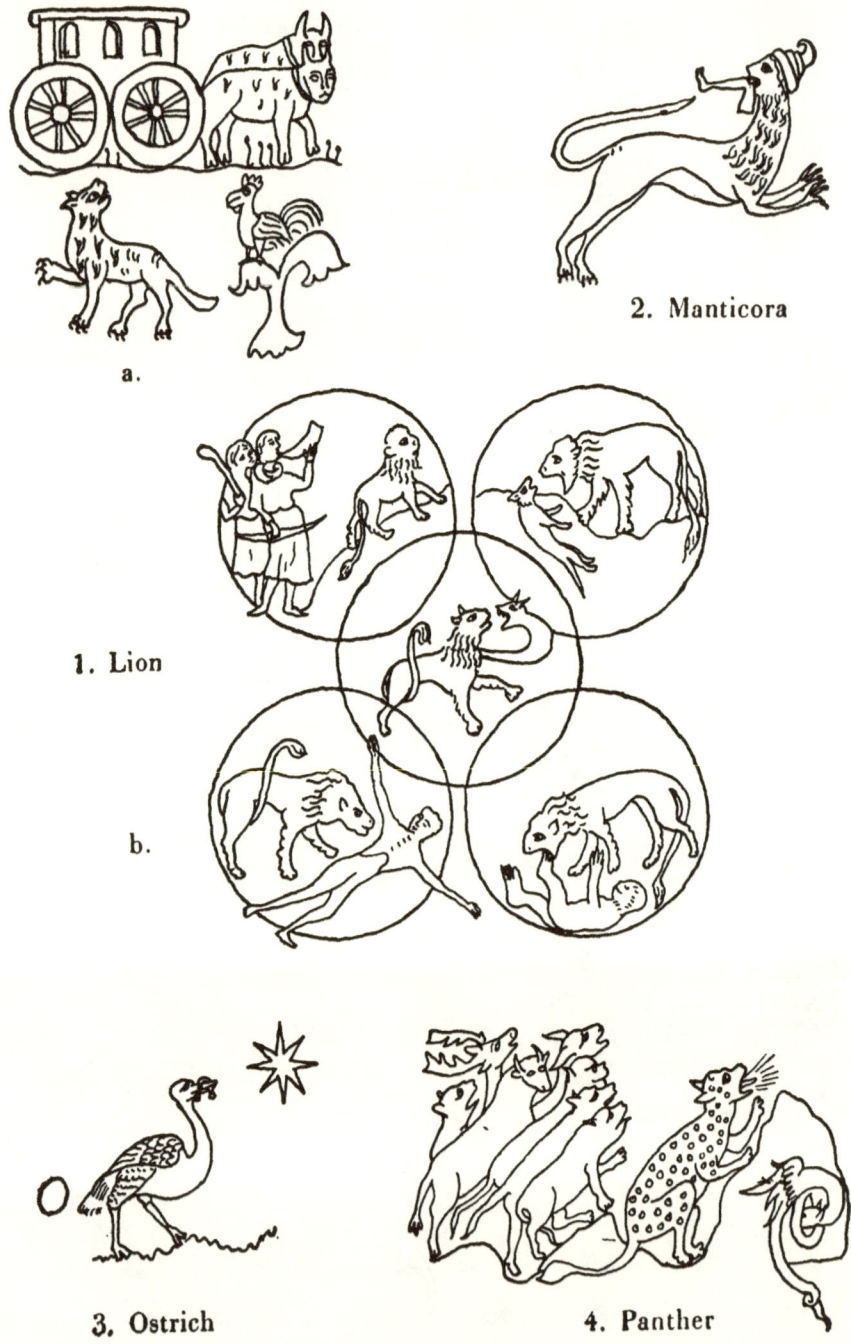

a.

2. Manticora

1. Lion

b.

3. Ostrich

4. Panther

PLATE VII

1. Phœnix

2. Onager

3. Salamander

4. Pelican   5. Peridexion Tree   6. Sawfish

PLATE VIII

a.

b.

1. Sawfish

2. Siren & Onocentaur

3. Siren

a.

b.

4. Snake   5. Stag   6. Viper

PLATE IX

a.

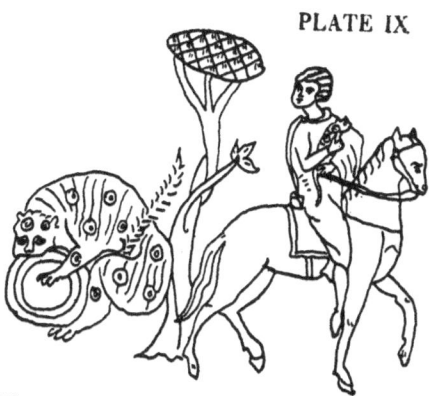

1. Tiger

b.

a.

c.

2. Unicorn

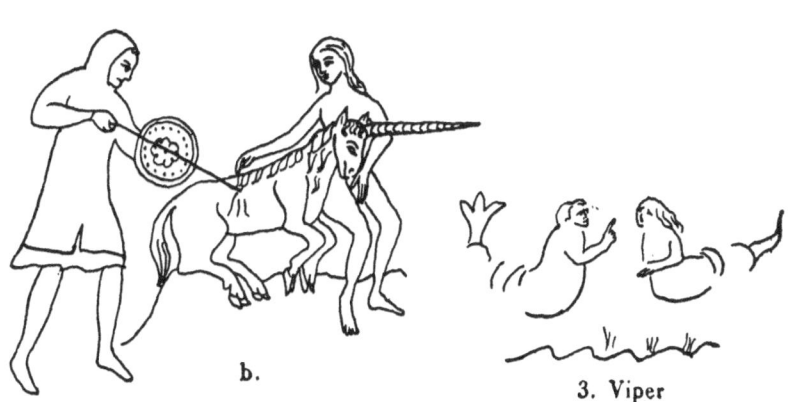

b.

3. Viper

PLATE X

1. Viper

2. Vulture

3. Weasel

4. Wolf

5. Yale

6. Woodpecker

APPENDIX

NON-BESTIARY MATERIAL IN PIERRE
DE BEAUVAIS

The following twelve descriptions found in Pierre de Beauvais'
long version of the *Bestiaire* have no close connection, apart from
the possible exception of the Echeneis, with any accounts found
in the traditional *Physiologus* or in the expanded form of the
bestiary such as the manuscripts of the Second or Third Family.
For the most part it has been impossible to find with any degree
of certainty a definite source for the various items; all that is
possible at present is to indicate surface similarities with contemporary Latin works, and occasionally even these have not been
discovered. Whether Pierre de Beauvais drew his additions from
one work which remains unknown or whether he dipped here and
there for the miscellaneous topics, either action is feasible but neither
is verifiable at this time.

ALERION.

alerion.

Although this bird is known in heraldry as a kind of eagle,[1] its
story in PB (II,162) is unlike that associated with the eagle apart
from its being credited with lordship over all other birds. In size
it is somewhat larger than an eagle, the color of fire, and with
wings sharp as a razor. Only one pair exists in the world. After
sixty years two eggs are laid, which are then brooded for sixty
days and nights. When the young are born, the parents, accompanied

---

[1] See Godefroy I,218.

by birds of the region, fly swiftly to the sea, plunge into the water, and drown. Returning to the young, the other birds guard and nourish them until they are strong and can fly.

An account similar to this is found in the French prose version of the letter of Prester John on the marvels of the East,[2] but its ultimate source has not been ascertained.

In Arsenal 3516, f. 202 the illustration shows two birds diving toward the sea while a single bird feeds the two young left in the nest.

ARGUS.

Argus le vachier.

Appearing only in PB (II,181) and some later French bestiaries is a picturesque account of the classical story of Argus. There was once a lady who had a cow very dear to her which she gave to a cowherd named Argus to guard. Of Argus' hundred eyes, two slept while the others watched. In spite of this vigilance the cow was lost because a man whom the cow had loved sent his son *Mercurius* to lull Argus to sleep with his pipe playing. Mercury then cut off the head of Argus and led the cow to his father.

In Alexander Neckam's *De naturis rerum* (i.39) appears a description not unlike this one of Argus.

Arsenal 3516, f. 203 portrays Mercury blowing on a horn while Argus, whose head is studded with eyes, listens. The miniature in Montpellier, H. 437, f. 214v. depicts Mercury leading off the cow after beheading Argus.

BARNACLE GOOSE.

bernekke, bernace; "l'arbre dont li oisel naisent fors et chient jus quant il sont meur", "des annes de la mer".[3]

Unique in French bestiaries is PB's account (II,216) of the Barnacle Goose, although it makes an unusual appearance in two

---

[2] Prester John's letter says that the pair lives forty years before laying the eggs. See *Oeuvres complètes de Rutebeuf*, ed. Achille Jubinal (Paris, 1839), II, 456.

[3] The first title is from Arsenal 3516 and the second from Vatican, Reg. 1323.

Latin bestiaries.[4] PB attributes to *Phisiologes* his information that
a tree growing over water bears birds resembling geese, only smaller.
While young the birds hang by their beaks, and when they mature,
they fall like ripe pears. Those falling on the water float and are
safe; those falling on land die.

This description appears more closely related to the one found
in Alexander Neckam's *De naturis rerum* (i.48), where the *bernekke*
are said to be born from wood exposed to salt water or from trees
planted on the edge of the shore, than to Giraldus Cambrensis'
notice, which only mentions the birth of the birds from deal planks.[5]

The illustration in PB (Pl. II, Fig. 3) is similar to those existing in the two Latin manuscripts except that they do not contain
the swimming birds.

BASILISK.

basilecoc, basilique.

PB's tale of the Basilisk (II,213-214) is different from the descriptions of this serpent that are contained in the enlarged Latin
bestiary, where its fantastic generation is not mentioned. According
to PB when the cock is seven years old, to its amazement an egg
is formed within it. After great suffering it secretly seeks a warm
place — a dung heap or a stable — where it digs a hole. Ten times
a day it goes there in hope of delivery. The toad smells the venom
inside the cock and watches carefully because it wants to hatch
the egg. This it does, and when the animal is born it has the upper
parts of a cock and the lower body of a snake. Other details are
included in PB's account. To avoid being seen the basilisk seeks
a crevice or a well, for if seen first by a man it dies; otherwise the
man is the victim. Since the basilisk casts poison from its eyes,
whoever wants to kill it should have a glass vessel, which stops
the poisonous glance and returns the venom to the basilisk.[6]

---

[4] B.M., Harl. 4751, f. 36 and Bodl. 764, f. 58v., where after the initial
sentence, "Sunt in Ybernia aves multe que bernace vocantur", the text of
Giraldus Cambrensis' *Topographica Hibernica* (i.15) is followed.

[5] For a brief history of the Barnacle Goose see P. Ansell Robin, *op. cit.*,
pp. 32-36.

[6] In PB's acount of the Yale (III,223), which he calls a *centicore*, the
basilisk is described as killing this animal by pricking it between the eyes,
though how the basilisk became connected with the *centicore* cannot be

An explanation of the basilisk's birth from a cock's egg has been proposed by P. Ansell Robin,[7] who thinks that this story probably had its origin in the Septuagint version of Isaiah 59:5, which, literally rendered, reads: "They break the eggs of adders and weave the spider's web: he who would eat of their eggs, having crushed the wind-egg (οὔριο), finds in it a basilisk." Aristotle (vi.559b 17) in discussing the wind-egg had spoken of substances resembling an egg being found in the cock. "It may therefore have been inferred that a basilisk was sometimes hatched from a cock's egg."[8] The cock and the basilisk are found linked in Aelian (iii.87).

The basilisk and toad appear together in Arsenal 3516, f. 204v.

CRICKET.

crisnon, gresillon.

According to PB (II, 155) the cricket so likes to sing that it loses its appetite, forgets everything, lets itself be hunted, and dies singing.

In the drawing of Arsenal 3516, f. 201 the cricket appears as a small, two-horned animal in front of a hole in the ground, whereas in Montpellier, H. 437, f. 206 there seem to be several crickets on a hearth.

ECHENEIS.

remora; essinus.

In the extensive chapter on the nature of various fish H (iii.55) includes the echeneis with a lengthy etymology which, unusual to say, is not found in Isidore (xii.6.34) but is alluded to in Pliny (ix.25.41). This is a small fish a half foot in length which takes its name from clinging to a ship and delaying it. The name is derived from the verb ἔγω, which is *habeo*, "to have", or ἔγομαι, which

---

explained. See George C. Druce, "Notes on the History of the Heraldic Jall or Yale", *op. cit.,* p. 186.

[7] The basilisk and the cockatrice are thoroughly investigated and their confused existence clearly presented by P. Ansell Robin, *op. cit.,* pp. 86-91 and Appendix.

[8] *Ibid.,* p. 87.

is *haereo*, "to adhere", and ναῦς, "boat". When the wind rushes or the storm rages, the boat seems to be rooted in the sea. The Latins call this fish *remora*, "delayer", because it forces vessels to remain still (*remoror*).

PB (IV,74) says that this fish of remarkable delaying ability is from the Indian Sea. Apparently PT's sawfish has assumed some of the traits of the echeneis, though by what means is not clear.

A fish swims under a ship in Arsenal 3516, f. 210v.

ELEMENTS and SENSES.

An unusual addition to the traditional bestiary material is first found in PB's chapter (IV,77) entitled "De quoi li home est fais et de sa nature", in which each of the four elements is represented by an animal, and the five senses are mentioned but not linked with any animal.[9] The salamander is said to live on fire, the chameleon — here called a bird — lives on wind, the herring on water, and the mole on earth. From its subterranean existence the mole is logically linked with the earth which Isidore (xii.3.5) says it eats, and from ancient times there was a belief that the chameleon lived on air alone (Pliny viii.33.51), but why the herring is used to symbolize the element of water, I have been unable to discover.

To portray the four elements in Arsenal 3516, f. 211, a naked man is drawn surrounded by a fish in water, a flying bird (the *gamalien*), a winged, four-footed salamander in fire, and a mole in the ground.

---

[9] In the *Bestiaire d'Amour* of Richard de Fournival (36.2-6) each sense is also identified with an animal (the plover is substituted for the chameleon in the elements); but in PB the only senses thus distinguished are found in the following chapter (IV,80), where the vulture is connected with the sense of smell and the *lieus* (or *liens*, Montpellier H. 437), a small white worm which can see through walls, is associated with sight. For a short examination of the "sensory champions" see H. W. Janson, *op. cit.*, pp. 239-40, and for hypotheses concerning the transformation of the lynx, noted in antiquity for its keen-sightedness, into a little worm, see John Holmberg (ed.), *Eine Mittelniederfrankische Übertragung des Bestiaire d'Amour* (Uppsala Universitet Årsskrift, 1925), pp. 243-44, and Edward B. Ham, "The Cambrai Bestiary", *Modern Philology*, XXXVI (1939), 229.

HARPY.

arpie.

PB's account of the harpy (II,157) is not based on the classical conception of the rapacious bird-women described by Homer and Virgil, but is similar to the description found in Odo of Cheriton.[10] According to PB, this beast resembles a horse and a man, with a lion's body, serpent's wings, and a horse's tail. Exceedingly cruel, it kills the first man it meets; then going to water and seeing its own reflection, it realizes that it has killed a fellow creature. It is grief-stricken, and its sadness is renewed each time the reflection is seen.

The drawing on Arsenal 3516, f. 201v. depicts a winged lion with a face perhaps human, standing beside water. Under the beast lies a dead man.

MUSCALIET.

muscaliet.

This unidentified animal — whose name has in all probability been incorrectly interpreted as the *musaraigne,* "shrew"[11] — is described by PB (IV,84) as living in the country of the three dry trees which announced to Alexander of Macedonia the day of his death. With body like a hare's but smaller, legs and tail like a squirrel's, it flies from branch to branch by the strength of its tail. Weasel-eared with a mole-like muzzle, its teeth resemble those of a boar and its hair that of a pig. This animal can quickly run up any tree, whose leaves or fruit it then devastates. Underground it hollows out the tree for its nest, where its hot nature causes the tree to dry up and die.

The *muscaliet* in a composite drawing in Arsenal 3516, f. 212 is a squirrel-like animal which burrows in the ground, emerges from a hole in a tree, and perches on top of a tree.

---

[10] See Léopold Hervieux, *Les Fabulistes latins* (Paris, 1884), II, 699.

[11] Oliver H. Prior (ed.), *L'Image du Monde de maître Gossouin* (Lausanne, 1913), glossary. The description of this animal hardly resembles the *musquelibet* noted by Albertus Magnus (*De animalibus,* lib. XXII, tract. 2, cap. 1.75), to which Cahier (IV,83) likens it. For the present it will apparently have to remain with the unknown animals found in PB.

ORPHAN BIRD.

orphanay (Vatican, Reg. 1323), *rafanay* (Phillipps 6739).

Only in PB (IV,85) does the description of this unusual bird appear and only in two manuscripts is any name given to it. So far no source has been found for this tale, although Cahier would link it with a remotely similar account in an Armenian bestiary of a bird called *Gérahav*.[12] The account in PB begins by locating the bird in India, where it inhabits a sea called "la mer darenoise".[13] It is crested, with a neck and chest like a peacock's, swan's feet, an eagle's beak, and the body of a crane with red, white, and black wings. As soon as its eggs are laid on the water, the young chick is within the shell, and the mother's intelligence senses which egg will contain the best offspring. If the chick is to be good, the egg will drop a little below the mother in the water, but that of the "deputaire pochin" falls to the sand on the sea bottom. At hatching time the good eggs rise under the mother's wing, where the young come forth and are joyfully led to their father. The bad eggs hatch on the floor of the sea, where the young live in darkness and grief.

A long-necked crested bird sits on the water in Arsenal 3516, f. 212v., while four large eggs float in the sea among fish.

TITMOUSE.

masenge.

In an account almost entirely free from the fabulous, PB (III,265) contains an elaborate description of the titmouse's appearance. The titmouse is said to have beautiful feathers of many colors. It has

---

[12] Charles Cahier, *Nouveaux mélanges d'archéologie* (Paris, 1874), I, 137.

[13] Cahier says concerning this word: "Si *darenoise* doit signifier quelque chose, ce pourrait bien être une altération de *tyrrhenum* (ou même *tyrium*) *mare*; quoiqu'il soit difficile de dire par où cette mer touche à l'Inde. ..." (IV, p. 85, n. 2). *Darenoise* might more plausibly, with the unexplained addition of an initial *d,* be derived from the adjective *arenosus,* "sandy", as it appears in the Latin and French letters of Prester John published by Jubinal, *op. cit.,* Vol. II. The first allusion speaks of "mare harenosum sine aqua" (p. 447) and the second, "la mere arenouse" (p. 458). Incidentally, the definition given by Tobler-Lommatzsch (I,513) for *arenos,* citing the above French passage, as "im fernen Osten", is incorrect.

a curious nature and wants to see other birds, whether dead or alive. When the hunter who has set a snare whistles and shows a bird, the titmouse flies toward the trap and is caught by its feet, as are many other birds who congregate upon hearing its cries.

WILD MAN.

ome sauvage.

This non-bestiary chapter of PB's (IV,76) has slightly different titles in the three main manuscripts in which it exists: Arsenal 3516, "Del sagetaire et del salvage home"; Montpellier H. 437, "Des gens as cornes ou front"; Vatican, Reg. 1323, "De l'ome sauvage". The account tells of the wild men dwelling in India who have a horn on their foreheads and who constantly fight against the *sagetaires*. To protect himself from wild animals which exist in abundance, the wild man prefers tree tops. He is also naked unless he happens to kill a lion and wear its skin.[14] Druce cites the use of the term "Physiologus" at the beginning of PB's chapter as evidence that the story must have come from an early Latin or Greek *Physiologus*,[15] but it is quite clear that PB used this term indiscriminately and that nothing can be proved by its presence. The origin of this short notice is unknown to me.

The illustrations of this chapter have a certain amusing aspect to them. In Arsenal 3516, f. 210v. a centaur shoots an arrow at a lion-robed, unicorned man who aims a spear at him, and in Montpellier H. 437, f. 248v. a man's head and a woman's head, each sporting a long horn, emerge from the top of a tree while on the ground prowls some sort of "vermine" or "sauvagine", to use PB's terms.

---

[14] Jubinal, *op. cit.,* II,458, quotes a similar description from a thirteenth century French translation of the letter to Prester John. This and PB's account of the wild man are unmentioned in Richard Bernheimer's *Wild Men in the Middle Ages* (Harvard University Press, 1952).

[15] George C. Druce, "Some Abnormalities...", *op. cit.,* p. 161.

BIBLIOGRAPHY

Aelianus, Claudius. *De natura animalium*. 2 vols. London, 1744.

Ahrens, Karl. *Buch der Naturgegenstände*. Kiel, 1892.

———. *Zur Geschichte des sogenannten Physiologus*. Ploen, 1885.

Ambrose, Saint. *Hexaemeron*, ed. C. Schenkl. Vol. XXXII, Part. I. *Corpus Scriptorum Ecclesiasticorum Latinorum*. Vienna, 1937.

Appel, Carl. *Provenzalische Chrestomathie*. Leipzig, 1895.

Aristotle. *Historia animalium*, trans. D'Arcy Wentworth Thompson. Oxford, 1910.

Barb, A. A. "Birds and Medical Magic: 1. The Eagle-Stone; 2. The Vulture Epistle", *Journal of the Warburg and Courtauld Institutes*, XIII (1950), 316-322.

Barbier de Montault, Xavier. "Fragments d'un Physiologus du XIIe siècle, à Monza", *Le Manuscrit*, II (1895), 181-184.

Beaunier, André. Review of Karviev, "Documents et remarques pour l'histoire littéraire du *Physiologus*", *Romania*, XXV (1896), 459-465.

Berger de Xivrey, Jules. *Traditions Tératologiques*. Paris, 1836.

Bernheimer, Richard. *Wild Men in the Middle Ages*. Cambridge, Mass., 1952.

Bestiaire d'amour rimé, ed. Arvid Thordstein. *Études romanes de Lund*, II (1941).

Brehaut, Ernest. "An Encyclopedist of the Dark Ages: Isidore of Seville", *Columbia University Studies in History, Economics and Public Law*, XLVIII (1912), 7-274.

Brunetto Latini. *Li Livres dou Trésor*, ed. Francis J. Carmody. Berkeley, Calif., 1948.

Cahier, Charles. *Nouveaux mélanges d'archéologie, d'histoire et de littérature*. Vol. I. Paris, 1874.

Carmody, Francis J. (ed.). *Physiologus Latinus: Éditions préliminaires versio B*. Paris, 1939.

———. (ed.). "Physiologus Latinus Versio Y", *University of California Publications in Classical Philology*, XII (1941), 95-134.

———. "De Bestiis et Aliis Rebus and the Latin Physiologus", *Speculum*, XIII (1938), 153-159.

———. "Quotations in the Latin Physiologus from Latin Bibles earlier than the Vulgate", *University of California Publications in Classical Philology*, XIII (1944-1950), 1-8.

———. "Le Diable des Bestiaires", *Cahiers de l'Association internationale des Études françaises*, Nos. 3-5 (Juillet 1953), 79-85.

206   BIBLIOGRAPHY

Carmody, Francis J., trans. *Physiologus, The very ancient book of beasts, plants and stones, translated from Greek and other languages*. San Francisco, 1953.

Cohn, Carl. *Geschichte des Einhorns*. Berlin, 1896.

Collins, A. H. "Some Twelfth-Century Animal Carvings and their Sources in the Bestiaries", *Connoisseur*, CVI (1940), 238-243.

Coulter, Cornelia C. "The 'Great Fish' in Ancient and Medieval Story", *Transactions of the American Philological Association*, LVII (1926).

Cronin, Grover. "The Bestiary and the Mediaeval Mind - Some Complexities", *Modern Language Quarterly*, II (1941), 191-198.

Delatte, Louis, ed. *Textes latins et vieux français relatifs aux Cyranides*. Fasc. XCIII. Bibliothèque de la Faculté de Philosophie et Lettres de l'Université de Liège. Paris, 1942.

Denis, Ferdinand. *Le Monde enchanté, cosmographie et histoire naturelle fantastiques du moyen âge*. Paris, 1843.

Druce, George C. "The Symbolism of the Crocodile in the Middle Ages", *Archaeological Journal*, LXVI (1909), 311-338.

——. "The Amphisbaena and its Connexions in Ecclesiastical Art and Architecture", *Archaeological Journal*, LXVII (1910), 285-317.

——. "Notes on the History of the Heraldic Jall or Yale", *Archaeological Journal*, LXVIII (1911), 173-199.

——. "The Caladrius and its Legend, Sculptured upon the Twelfth-Century Doorway of Alne Church, Yorkshire", *Archaeological Journal*, LXIX (1912), 381-416.

——. "Some Abnormal and Composite Human Forms in English Church Architecture", *Archaeological Journal*, LXXII (1915), 135-186.

——. "The Elephant in Medieval Legend and Art", *Archaeological Journal*, LXXVI (1919), 1-73.

——. "Legend of the Serra or Saw-Fish", *Proceedings of the Society of Antiquaries of London*, 2nd Series, XXXI (1919), 20-35.

——. "The Mediaeval Bestiaries, and their Influence on Ecclesiastical Decorative Art", *British Archaeological Journal*, New Series, XXV (1919), 41-82; XXVI (1920), 35-79.

——. "An Account of the Μυρμηχολέων or Ant-Lion", *Antiquaries Journal*, III (1923), 347-364.

——., trans. *The Bestiary of Guillaume le Clerc*. Printed for private circulation by Headley Brothers, Invicta Press, 1936.

Duncan, Thomas S. "The Weasel in Religion, Myth and Superstition". *Washington University Studies* (Humanistic Series), XII (1924), 33-66.

Epiphanius, *Physiologus*. Migne, *Patr. Gr.*, XLIII, Cols. 518-534.

Faral, Edmond. "La queue de poisson des sirènes", *Romania, LXXIV* (1953), 433-506.

Fournival, Richard de. *Li Bestiaires d'Amours di Maistre Richart de Fornival e li Response du Bestiaire*, ed. Cesare Segre. Milan, 1957.

Garver, Milton S. "Sources of the Beast Similies in the Italian Lyric of the Thirteenth Century", *Romanische Forschungen*, XXI (1905-08), 276-320.

——. "Supplementary Italian Bestiary Chapters", *Romanic Review*, XI (1920), 308-327.

——, and McKenzie, K. *Il Bestiario Toscano*. Rome, 1912.

Gervaise. "Le Bestiaire de Gervaise", ed. Paul Meyer. *Romania*, I (1872), 420-443.

Goldstaub, Max. "Der Physiologus und seine Weiterbildung", *Philologus,* Supplementband VIII (1899-1901), 339-404.

――――. "Physiologus-Fabelein über das Brüten des Vogels Strauss", *Fest-schrift Adolf Tobler.* Braunschweig, 1905. Pp. 153-190.

――――, and Wendriner, Richard. *Ein Tosco-Venezianischer Bestiarius.* Halle, 1892.

Guillaume le Clerc. *Le Bestiaire, das Thierbuch des normannischen Dichters Guillaume le Clerc,* ed. Robert Reinsch. Leipzig, 1890.

Halliday, W. R. "Picus-who-is-also-Zeus", *Classical Review,* XXXVI (1922), 110-112.

Ham, Edward B. (ed.). "The Cambrai Bestiary", *Modern Philology,* XXXVI (1939), 225-237.

Hamilton, G. L. Review of Ch.-V. Langlois, *La Connaissance de la nature et du monde* (2nd ed., Paris, 1927), *Speculum,* IV (1929), 110-116.

Hastings, James (ed.). *A Dictionary of the Bible.* 5 vols. New York, 1902.

Heckscher, William S. "Bernini's Elephant and Obelisk", *Art Bulletin,* XXIX (1947), 155-182.

Heider, Gustav. "Physiologus nach einer Handschrift des XI. Jahrhunderts", *Archiv für Kunde österreichischer Geschichts-Quellen.* Dritter Jahrgang, Zweiter Band. Vienna, 1850. Pp. 541-582.

Henkin, Leo J. "The Carbuncle in the Adder's Head", *Modern Language Notes,* LVIII (1943), 34-39.

Hermannsson, Halldor. Introduction. "The Icelandic Physiologus", *Icelandica,* XXVII (1938).

Hilka, Alfons. "Die anglo-normannische Versversion des Briefes des Presby-ters Johannes", *Zeitschrift für Französische Sprache und Litteratur,* XLIII (1915), 82-112.

Höhna, Heinrich. *Der Physiologus in der elisabethanischen Literatur.* Erlan-gen, 1930.

Hommel, Fritz. *Die Aethiopische Uebersetzung des Physiologus.* Leipzig, 1877.

―――― (ed.). "Der äthiopische Physiologus", *Romanische Forschungen,* V (1890), 13-36.

Horapollo. *The Hieroglyphics of Horapollo.* Trans. George Boas. New York, 1950.

Hubaux, Jean, and Leroy, Maxime. *Le Mythe du phénix dans les littératures grecque et latine.* Fasc. LXXXII. Bibliothèque de la Faculté de Philoso-phie et Lettres de l'Université de Liège. Liège and Paris, 1939.

Hugo of Saint Victor. *De bestiis et aliis rebus.* Migne, *Patr. Lat.,* CLXXVII. Cols. 15-164.

Hulme, Edward F. *Natural History Lore and Legend.* London, 1895.

L'Image du monde de maître Gossouin. Ed. Oliver Prior. Lausanne, 1913.

Isidore of Seville. *Etymologiarum sive originum Libri XX.* Ed. W. M. Lind-say. Oxford, 1911.

Ives, Samuel, and Lehmann-Haupt, Hellmut. *An English 13th Century Bes-tiary.* New York, 1942.

James, Montague Rhodes. *The Bestiary.* Oxford, 1928.

――――. "The Bestiary", *History,* New Series, XVI, No. 61 (April, 1931), 1-11.

――――. "The Bestiary in the University Library", *Aberdeen University Library Bulletin,* No. 36 (January, 1928), 1-3.

――――. *Peterborough Psalter and Bestiary.* Oxford, 1921.

Janson, Horts W. *Apes and Ape Lore in the Middle Ages and the Renaissance.* London, 1952.

Karajan, Theodor. *Deutsche Sprach-Denkmale des zwölften Jahrhunderts.* Vienna, 1846.

Keller, Otto. *Die antike Tierwelt.* 2 vols. Leipzig, 1909.

Konstantinova, Alexandra. "Ein englisches Bestiar des zwölften Jahrhunderts", *Kunstwissenschaftliche Studien,* IV (1929).

Krappe, Alexander: "The Historical Background of Philippe de Thaün's *Bestiaire*", *Modern Language Notes,* LIX (1944), 325-327.

Kroll, Wilhelm. "Bolos und Demokritos", *Hermes,* LXIX (1934), 228-232.

Lactantius. *De ave phoenice.* Ed. Mary C. Fitzpatrick. Published Ph. D. dissertation. University of Pennsylvania, 1933.

Land, J. P. N. "Physiologus Leidensis", and "Scholia in Physiologum Leidensem", *Anecdota Syriaca,* IV (1875), 31-98, 115-176.

Langlois, Ch.-V. *La Connaissance de la nature et du monde.* Vol. III of *La Vie en France au moyen âge.* 2nd ed. Paris, 1927.

Van Lantschoot, Arn. "A propos du Physiologus", Reprinted from *Coptic Studies in Honor of Walter Ewing Crum. Byzantine Institute Bulletin,* No. 2, 1950. Pp. 339-363.

Lauchert, Friedrich. *Geschichte des Physiologus.* Strassburg, 1889.

———. "Zum Physiologus: Der tiergeschichtliche Abschnitt der Acerba des Cecco d'Ascoli, eine Bearbeitung des Physiologus", *Romanische Forschungen,* V (1890), 1-12.

———. "Der Einfluss des Physiologus auf den Euphuismus", *Englische Studien,* XIV (1890), 188-210.

Laurent, M. "Le phénix, les serpents et les aromates dans une miniature du XIIᵉ siècle", *L'Antiquité classique,* IV (1935), 375-401.

Libellus de Natura Animalium. Reproduced in facsimile with an introduction by J. I. Davis. London, 1958.

Lüdtke, Willy. "Zum armenischen und lateinischen Physiologus", *Huschardzan Festschrift.* Vienna, 1911.

McCulloch, Florence. "Pierre de Beauvais' *Lacovie*", *Modern Language Notes,* LXXI (1956), 100-101.

———. "The Metamorphoses of the Asp", *Studies in Philology,* LVI (1959), 7-14.

McKenzie, Kenneth. "Unpublished Manuscripts of Italian Bestiaries", *Publications of the Modern Language Association,* XX (1905), 380-433.

Mai, Angelo. "Exerpta ex Physiologo", *Classici Auctores,* Vol. VII, Rome, 1835. Pp. 589-596.

Manitius, Max. *Geschichte der Lateinischen Literatur des Mittelalters.* 3 vols. Munich, 1911-1931.

Mann, Max Friedrich. "Der Bestiaire Divin des Guillaume le Clerc", *Französische Studien,* VI Band, 2 Heft (1888), 1-106.

———. "Der Physiologus des Philipp von Thaün und seine Quellen", *Anglia,* VII (1884), 420-468; IX (1886), 391-434, 447-450.

———. Review of Friedrich Lauchert. *Geschichte des Physiologus. Englische Studien,* XIV (1890), 123-127, 297-299.

de Mély, F. and Ruelle, Ch.-Em. *Les lapidaires de l'antiquité et du moyen âge.* Tome II, *Les lapidaires grecs.* Paris, 1898.

Menhardt, Hermann. "Der Millstätter Physiologus und seine Verwandten", *Kärntner Museumsschriften,* XIV (1956).

Menhardt, Hermann. "Der Physiologus im Schloss Tirol", *Der Schlern*, XXXI (1957), 401-405.

Meyer, Paul. "Les Bestiaires", *Histoire littéraire de la France*, XXXIV (1914), 362-390.

Millar, Eric. *A Thirteenth Century Bestiary in the Library of Alnwick Castle.* Oxford, 1958.

Morson, John. "The English Cistercians and the Bestiary", *Bulletin of the John Rylands Library Manchester*, XXXIX (1956), 146-170.

Neckam, Alexander. *De naturis rerum.* Ed. Thomas Wright. Rolls Series. London, 1863.

Old English Elene, Phoenix, and Physiologus, ed. Albert Stanburrough Cook. New Haven, 1919.

Perry, Ben. E. "Physiologus", Pauly-Wissowa, *Real-Encyclopädie der classischen Altertumswissenschaft.* Neue Bearbeitung. XX (1950), 1074-1129.

——. Review of Francesco Sbordone. *Physiologus.* Milan, 1936. *American Journal of Philology*, LVIII (1937), 488-496.

Peters, Emil., trans. *Der Physiologus aus dem griechischen Original übertragen.* Munich, 1921.

Philippe de Thaon. *Le Bestiaire de Philippe de Thaün*, ed. Emmanuel Walberg. Paris and Lund, 1900.

Pierre de Beauvais, "Bestiaire en prose de Pierre le Picard", ed. Charles Cahier and Arthur Martin. *Mélanges d'archéologie, d'histoire et de littérature.* Paris. Vol. II (1851), pp. 85-100, 106-232; Vol. III (1853), pp. 203-288; Vol. IV (1856), pp. 55-87.

Pitra, J. B. *Spicilegium Solesmense.* Vol. III. Paris, 1855.

Pliny. *Naturalis historiae libri XXXVII*, Ed. C. Mayhoff. 5 vols. Leipzig, 1892-1909.

Plutarch. *Moralia.* Ed. G. Bernardakis. 6 vols. Leipzig, 1888-1896.

Queen Mary's Psalter. Introduction by Sir George Warner. London, 1912.

Randolph, Charles B. "The Mandragora of the Ancients in Folk-Lore and Medicine", *Proceedings of the American Academy of Arts and Sciences*, XL (1905), 487-537.

Raynaud, Gaston (ed.). "Poème moralisé sur les propriétés des choses", *Romania*, XIV (1885), 442-484.

Robin, P. Ansell. *Animal Lore in English Literature.* London, 1932.

Rose, William. *The Epic of the Beast (Physiologus*, trans. James Carlill). London, n. d.

Sbordone, Francesco (ed.), *Physiologus.* Milan, 1936.

——. *Ricerche sulle fonti e sulla composizione del Physiologus greco.* Naples, 1936.

——. "La Tradizione manoscritta del *Physiologus* Latino", *Athenaeum*, Nuova Serie, XXVII (1949), 246-280.

Shepard, Odell. *The Lore of the Unicorn.* London, 1930.

Solalinde, A. G. "El *Physiologus* en la *General Estoria* de Alfonso X", *Mélanges d'histoire littéraire générale et comparée offerts à Fernand Baldensperger.* Paris, 1930.

Solinus. *Collectanea rerum memorabilium*, ed. th. Mommsen. Berlin, 1864.

——. *The Excellent and Pleasant Worke of Caius Julius Solinus.* Trans. Arthur Golding. Gainesville, Florida, 1955.

Spitzer, Leo. "Auf keinen grünen Zweig kommen", *Modern Language Notes*, LXIX (1954), 270-273.

Stettiner, Richard. *Die illustrierten Prudentiushandschriften.* Tafelband. Berlin, 1905.
Strzygowski, Josef. "Der Bilderkreis des griechischen Physiologus", *Byzantinisches Archiv.* Heft 2, 1899, Pp. 1-130.
Theobaldus. *Physiologus of Theobaldus.* Ed. Richard Morris. (*An Old English Miscellany,* Early English Text Society). London, 1872. Pp. 201-209.
———. *Physiologus a Metrical Bestiary of Twelve Chapters by Bishop Theobald,* trans. Alan W. Rendell. London, 1928.
Thompson, C. J. S. *The Mystic Mandrake.* London, 1934.
Thompson, D'Arcy Wentworth. *A Glossary of Greek Birds.* Oxford, 1936.
Thorndike, Lynn. *A. History of Magic and Experimental Science.* Vols. I, II. New York, 1929.
Tobler, Adolf. "Lateinische Beispielsammlung mit Bildern", *Zeitschrift für Romanische Philologie,* XII (1888), 57-88.
Tselos, Dimitri. "A Greco-Italian School of Illuminators and Fresco Painters: Its Relation to the Principal Reims Manuscripts and to the Greek Frescoes in Rome and Castelseprio", *Art Bulletin,* XXXVIII (1956), 1-30.
Voigt, Ernst. Review of Friedrich Lauchert. *Geschichte des Physiologus.* Strassburg, 1889. *Zeitschrift für Deutsche Philologie,* XXII (1890), 237-242.
"Der waldensische Physiologus", ed. Alfons Mayer. *Romanische Forschungen,* V (1890), 392-418.
Wellmann, Max. "Der Physiologos: ein religionsgeschichtlich-naturwissenschaftliche Untersuchung", *Philologus, Supplementband* XXII, Heft I (1930), 1-116.
White, Lynn Jr. "Natural Science and Naturalistic Art in the Middle Ages", *American Historical Review,* LII (1947), 421-435.
White, T. H. *The Book of Beasts.* London, 1954.
Wilhelm, Friedrich (ed.). "Dicta Chrysostomi", *Münchener Texte,* Heft 8 B (Kommentar), 1916. Pp. 13-52.
Woodruff, Helen. "Illustrated Manuscripts of Prudentius", *Art Studies,* VII (1929), 33-79.
———. "The Physiologus of Bern", *Art. Bulletin,* XII (1930), 226-253.

THIRD PRINTING
ADDITIONAL BIBLIOGRAPHY

Chapter I
Der Physiologus. Translated by Otto Seel. Zürich and Stuttgart, 1960.

Chapter II
Der altdeutsche Physiologus. Die Millstätter Reimfassung und die Wiener Prosa (nebst dem lateinischen Text und dem althoch deutschen Phsyiologus). Edited by Friedrich Maurer. Tübingen, 1967.
de Clercq, Carlo. "Le Rôle de l'image dans un manuscrit médiéval," *Gutenberg Jahrbuch,* 1962, 23-30.
Leclercq, J., "Les Peintures de la Bible de Morimondo," *Scriptorium,* X (1956), 22-26.
Lyna, F., "Een verluchte Engelsche bestiaris te Brussel (Hs. 8327-42)," *Gentsche Bijdragen tot de Kunstgeschiedenis,* IV (1939-40), 231-246.

BIBLIOGRAPHY 211

Menhardt, Hermann, "Die Zweiheit Genesis-Physiologus und der Zeitan-
satz der Exodus," *Zeitschrift für deutsches Altertum,* 1959, 257-271.
Physiologus; en bok om naturens ting. Translated by B. S. Nordin-Petterson.
Stockholm, 1957.
Physiologus Bernensis. Voll-Faksimile-Ausbage des Codex Bongarsianus
318 der Burgerbibliothek Bern. Commentary by Christoph von Steiger
and Otto Homburger. Basel, 1964.
Robinson, Margaret W. *Fictitious Beasts: a Bibliography,* London, 1961.
Silvestre, Hubert. "Enfin un manuscrit anglais du *De vita et moribus phi-
losophorum de Walter Burley* (Bruxelles, Bibl. Roy, 8327-42)," *Scriptori-
um,* XIII (1959), 255-259.
Theobaldus. "The Book as an Artist's medium: Theobald's Bestiary,"
Publishers Weekly, Vol. 187, No. 5 (February, 1965), 114-120.

Chapter III
Berkey, Max L., "Pierre de Beauvais: An Introduction to His Works,"
Romance Philology, XVIII (1965), 387-398.
Bestiaris. Edited by Saverio Panunzio. 2 vols. Barcelona, 1963, 1964.
McCulloch, Florence. "Pierre Gringore's *Menus Propos des amoureux* and
Richard de Fournival's *Bestiaire d'Amour,*" *Romance Notes,* X (Au-
tumn, 1968), 150-159.
———. "The Waldensian Bestiary and the *Libellus de Natura Animalium,*"
Medievalia et Humanistica, 15 (1963), 15-30.
Mermier, Guy. "The Bestiary of Gervaise," *Papers of the Michigan Acad-
emy of Science, Arts, and Letters,* 53 (1967), 337-352.
———. "De Pierre de Beauvais et particulièrement de son bestiaire: vers
une solution des problèmes," *Romanische Forschungen,* 78 (1966), 338-
371.
Radicula, Carla. "Il *Bestiaire d'Amours* capostipite di Bestiari latini e
romanzi," *Studi medievali,* 3ª Serie, III (1962), 576-606.
Smeets, J. R. "L'ordre des 'Animaux' dans le *Physiologus* de Philippe de
Thaün et la prétendue préséance de la perdrix sur l'aigle," *Revue Belge
de Philologie et d'Histoire,* 40 (1962), 798-803.

Chapter V
Gerhardt, Mia I. "The ant-lion," *Vivarium,* III (1965), 1-23.
Graham, Victor E. "The Pelican as Image and Symbol," *Revue de Lit-
térature comparée,* 36 (1962), 233-243.
Hubert, M. "La taille de la licorne," *Archivum latinitatis medii aevi,* XXVII
(1957), 167-187.
Koch, R. A. "The Salamander in Van der Goes' Garden of Eden," *Journal
of the Warburg and Courtauld Institute,* 28 (1965), 323-326.
MacKinney, Loren C. "Moon-Happy Apes, Monkeys and Baboons," *Isis,*
54 (1963), 120-122.
McCulloch, Florence. "Mermecolion—A Mediaeval Latin Word for 'Pearl
Oyster'," *Mediaeval Studies,* XXVII (1965), 331-334.
———. "Le Tigre au miroir: la vie d'une image de Pline à Pierre Gringore,"
Revue des Sciences humaines, 130 (1968) 149-160.
Menhardt, Hermann. "Die Mandragora im Millstätter Physiologus,"
Festschrift für Ludwig Wolff. Edited by Werner Schröder. Neumünster,
1962.

Thibout, Marc. "L'Eléphant dans la sculpture romane française," *Bulletin Monumental,* 1947, 183–195.

Varty, Kenneth. *Reynard the Fox.* Leicester, 1967.

Appendix

Coonen, L. P. "Medieval Skepticism of the Barnacle Goose Legend," *Turtox News,* Vol. 47, No. 6 (1969), 210–212.

George, W. "The Yale," *Journal of the Warburg and Courtauld Institute,* 31 (1968), 423–428.

McCulloch, Florence. "L'éale et la centicore: deux bêtes fabuleuses," *Mélanges offerts à René Crozet.* Poitiers, 1966. Vol. II, 1167–1172.

ADDENDA AND CORRIGENDA

1. *p. 31. Manuscripts of this text are also often listed in catalogues under the title *De tribus columbis*.

2. *p. 31. As a result of more recent research the following information should be added to note 32:

    A recent denial of Hugo's authorship of the *De bestiis* is found in Roger Baron, *Science et Sagesse chez Hugues de Saint-Victor* (Paris, 1957), pp. xxxi, xxxii. The rejection of the attribution of Book III to Guillelmus Peraldus (Guillaume Peyraut) is repeated by Hubert Silvestre, "Apropos du *Liber tertius* du *De bestiis et aliis rebus* et d'un passage des *Etymologiae* (11,2,33) d'Isidore de Seville," *Le Moyen Age*, 4e serie, 4-5 (1949), 247. The connection between Hugo of Saint Victor and the hybrid composition entitled the *De bestiis* is doubtless explained by the fact that the author of the allegorical treatise on birds, the *Aviarium* (which as a separate work ends with the chapter on Aquila (56), and which when followed directly by the treatise on animals, as in Migne's edition, ends with Fulica), was Hugo of Folieto, some of whose writings already in the thirteenth century were ascribed to the more famous Hugo. See Henri Peltier, "Hugues de Fouilloy, chanoine regulier, prieur de Saint-Laurent-au-bois," *Revue du moyen âge latin*, II (1946), 36-42.

3. *p. 32. To this list the following two illustrated manuscripts should be added:

    B.N., lat. 2495 A, f. 17-23. End XII cent. (*Avarium*, f. 1-16v.).
    B.N., lat. 2495 B, f. 29v.-47. XIII cent. (*Avarium*, f. 1-29).

4. *p. 34. Photographs of this manuscript, whose marginal animal drawings (f. 9-64) generally follow the order of Morgan 81, are in the New York Public Library (*ISM). These illustrations are briefly described by Donald Drew Egbert, *The Tickhill Psalter and Related Manuscripts* (New York and Princeton, 1940), pp. 166-68.

ADDENDA AND CORRIGENDA

5. *p. 44. Numerous allusions in the works of Peter Damien, the celebrated eleventh century ecclesiastical reformer, show the symbolical relations that can be created in the world of nature for the moral edification of man by a writer versed in traditional animal lore (perhaps derived in part from Ambrose's *Hexaemeron*). See especially his letter entitled *De bono religiosi status et variarum animantium tropologia* (Migne, *Patr. Lat.,* CXLV, Cols. 763-792) where he compares the abbey of Monte Cassino to Noah's ark and the animals found therein. See also Dom Jean Leclercq, *Saint Pierre Damien ermite et homme d'église* (Rome, 1960), pp. 177-191.

6. *p. 47. Not one, as originally stated, but two manuscripts exist of the Waldensian Bestiary: the Dublin manuscript and another at the University Library, Cambridge, Dd. 15.29, f. 17-49. See Mario Esposito, "Sur quelques manuscrits de l'ancienne littérature religieuse des Vaudois du Piémont, *"Revue d'Histoire ecclésiastique*, XLVI, N. 1-2 (1951), 148-49.

7. *p. 66. Monsieur Jodogne, in a personal communication, more convincingly proposes that *cuers* is a scribal misreading of *envers* in the original text which doubtless read *Philipon envers qui service* . . . .

8. *p. 71. Recently the style of an eleventh century Greek manuscript of the *Physiologus* (Milan, Ambrosiana E.16 Sup. Restaurato N. 83, f. 1-41), whose illustrations are characterized as "primitive," has been analyzed by Maria Luisa Gengaro, "A proposito delle inedite illustrazioni del Phisiologus greco della Biblioteca Ambrosiana," *Arte Lombarda*, Anno III, Numero I (1958), 19-27.

9. *p. 103. This scene is based on a verse from Genesis (49:17): "Fiat Dan coluber in via, cerastes in semita, mordens ungulas equi, ut cadat ascensor eius retro."

10. *p. 157. The peculiarity of the blood issuing from the pelican's mouth is probably a transference of the essential detail of the normal scene where though the blood gushes from the bird's side wounded by the beak, it could appear to come from the beak itself.